LIPIDS

Volume 1 / Biochemistry

LIPIDS

Volume 1 / Biochemistry

Editors

Rodolfo Paoletti
*Institute of Pharmacology
and Pharmacognosy
University of Milan
Milan, Italy*

Giuseppe Porcellati
*Director, Institute of Biological
Chemistry
University of Perugia,
Perugia, Italy*

Giovanni Jacini
*Experimental Station on
Oils and Fats
Milan, Italy*

Raven Press ■ New York

Made in the United States of America

International Standard Book Number 0–89004–028–1
Library of Congress Catalog Card Number 74–21982

Preface

The invited lectures and symposia presented at the International Congress for Fat Research held in Milan, Italy in September, 1974 are presented in two volumes.

The contributions that cover the biochemical aspects of Lipids are included in Volume 1. Of particular interest is the group of papers concerning phospholipases. These enzymes, believed to be the rate-limiting step for prostaglandin biosynthesis in mammalian systems, are discussed in considerable detail for the first time with a distinct interdisciplinary approach. The physical properties of biological membranes as related to their lipid composition and the newly discovered glycolipids and bacterial lipids complete this section of the proceedings. Another area, neglected until very recently, is the role of dietary lipids in postnatal development, particularly of the brain structures. Several authoritative papers deal with this new area of biochemical research.

Volume 2, which is devoted to lipid technology, presents numerous contributions on new developments in analytic methods with particular emphasis on NMR spectrometry, deuterium magnetic resonance, and gas chromatography combined with mass spectrometry. These technical developments have rapidly increased the information available on the role played by lipids in food, cosmetics, and drug preparations. Of particular interest in the section devoted to lipids in foodstuffs are several experimental contributions on olive oil as compared with other vegetable oils. The unique properties of this Mediterranean edible oil are discussed for the first time. Other application of lipids notably in drugs and cosmetics are reviewed. The important problems concerning possible toxic effects of cosmetic preparations and the absorption of pharmacologically active compounds through the skin make this volume very timely.

The two volumes should represent useful companions for the everyday work of lipid biochemists and technologists and for the increasing number of nonspecialists involved with the complex field of basic research and practical applications of lipids.

Rodolfo Paoletti
Giuseppe Porcellati
Giovanni Jacini
(July, 1975)

Contents of Volume 1

Lipases, Phospholipases, Lipoperoxidases, and Lipoxygenases

Physical Properties and Lipid Composition of Membranes

Lipids and Foodstuffs

Newly Discovered Lipids

(Cumulative index for both volumes appears in Volume 2, p. 539)

Contents of Volume 2

Advances in Surface-Active Agents

Flavors

Lipids in Drugs and Cosmetics

New Trends in Margarine and Hydrogenated Fats

Contributors

E. Agradi
Institute of Pharmacology and Pharmacognosy
University of Milan
Via A. Del Sarto 21
20129 Milan, Italy

B. Akesson
Department of Physiological Chemistry
University of Lund
P. O. Box 750
S-220 07
Lund 7, Sweden

Marta I. Aveldano
Instituto de Investigaciones Bioquimicas
Universidad Nacional del Sur
Bahia Blanca, Argentina

J. Barwick
Laboratory of Biochemistry
State University of Utrecht
Utrecht, Netherlands

Nicolas Bazan
Instituto de Investigaciones Bioquimicas
Universidad Nacional del Sur
Bahia Blanca, Argentina

B. Berra
Department of Biological Chemistry
University of Milan
School of Medicine
Via Saldini 50
20133 Milan, Italy

Luciano Binaglia
Istituto di Chimica Biologica
Università di Perugia
Facolta Medica
C.P. n. 3
06100 Perugia, Italy

J. Boldingh
Organic Chemistry Department
State University of Utrecht
Utrecht, The Netherlands

H. van den Bosch
Laboratory of Biochemistry
State University of Utrecht
Transitorium 3
Padualaan 8
Utrecht, The Netherlands

Eliahu Caspi
Worcester Foundation for Experimental Biology
222 Maple Avenue
Shrewsbury, Massachusetts 01545

M. Ciampa
Department of Biological Chemistry
University of Milan
School of Medicine
Via Saldini 50
20133 Milan, Italy

A. Colbeau
DRF/Biochimie
CEN-G., B.P. 85
38041, Grenoble-Cedex, France

F. Cuault
DRF/Biochimie
CEN-G., B.P. 85
38041, Grenoble-Cedex, France

G. Debernardi
Department of Biological Chemistry
University of Milan
School of Medicine
'Via Saldini 50
20133 Milan, Italy

H. J. Fallon
Virginia Commonwealth University
Medical College of Virginia
Richmond, Virginia 23298

Flaminio Fidanza
Istituto di Scienza dell' Alimentazione
Università degli Studi
016100 Perugia, Italy

Werner Fischer
Institute of Physiological Chemistry
University Erlangen-Nürnberg
Wasserturmstrasse 5
D-8520 Erlangen
West Germany

Kuo-Lan Fong
Biomembrane Research Laboratory
Oklahoma Medical Research Foundation
825 NE 13th Street
Oklahoma City, Oklahoma 73104

C. Galli
Institute of Pharmacology and Pharmacog-
  nosy
University of Milan
Via A. Del Sarto 21
20129 Milan, Italy

Norma M. Giusto
Instituto de Investigaciones Bioquimicas
Universidad Nacional del Sur
Bahia Blanca, Argentina

Howard Goldfine
Department of Microbiology
University of Pennsylvania
School of Medicine
Philadelphia, Pennsylvania 19174

Gianfrancesco Goracci
Istituto di Chimica Biologica
Università di Perugia
Facolta Medica
C.P. n. 3
06100 Perugia, Italy

L. Gosselin
University of Liège
Laboratory of Medical Chemistry
151 Boulevard de la Constitution
B-4000 Liège, Belgium

G. H. de Haas
Laboratory of Biochemistry
State University of Utrecht
Transitorium 3
Padualaan 8, Utrecht
The Netherlands

Ralph T. Holman
The Hormel Institute
University of Minnesota
Austin, Minnesota 55912

K. Roger Hornbrook
Department of Pharmacology
College of Medicine
University of Oklahoma Health Science
  Center
P. O. Box 26901
Oklahoma City, Oklahoma 73190

G. Jacini
Experimental Station of Oil and Fat
via Guiseppe Colombo 79
20133 Milan, Italy

A. T. James
Unilever Research
Colworth House
Sharnbrook, Bedford, England

R. Jeffcoat
Unilever Research
Colworth House
Sharnbrook, Bedford, England

Robert G. Jensen
Department of Nutritional Sciences
University of Connecticut
Storrs, Connecticut 06268

J. G. N. de Jong
Laboratory of Biochemistry
State University of Utrecht
Transitorium 3
Padualaan 8, Utrecht
The Netherlands

Hideo Kanoh
Department of Biochemistry
Sapporo Medical College
West-17, South-1
Sapporo 060 Hokkaido
Japan

M. Kates
Department of Biochemistry
University of Ottawa
Ottawa K1N 6N5, Canada

G. K. Khuller
Department of Microbiology
University of Pennsylvania
School of Medicine
Philadelphia, Pennsylvania 19174

Margaret King
Biomembrane Research Laboratory
Oklahoma Medical Research Foundation
825 NE 13th Street
Oklahoma City, Oklahoma 73104

B. de Kruyff
Department of Biochemistry
State University of Utrecht
Padualaan 8
De Uithof, Utrecht
The Netherlands

A. Kuksis
Banting and Best Department of Medical Research
University of Toronto
Toronto M5G 1LG, Canada

S. C. Kushwaha
Department of Biochemistry
University of Ottawa
Ottawa K1N 6N5, Canada

Edward Lai
Biomembrane Research Laboratory
Oklahoma Medical Research Foundation
825 NE 13th Street
Oklahoma City, Oklahoma 73104

R. G. Lamb
Laboratory of Biochemistry
State University of Utrecht
Padualaan 8, Utrecht
The Netherlands

Donald R. Lueking
Department of Medicine
University of Pennsylvania
Philadelphia, Pennsylvania 19174

M. Manto
Department of Biological Chemistry
University of Milan
School of Medicine
Via Saldini 50
20133 Milan, Italy

Paul B. McCay
Biomembrane Research Laboratory
Oklahoma Medical Research Foundation
825 NE 13th Street
Oklahoma City, Oklahoma 73104

James F. Mead
Laboratory of Nuclear Medicine and Radiation Biology
900 Veteran Avenue
University of California
Los Angeles, California 90024

J. J. Myher
Banting and Best Department of Medical Research
University of Toronto
Toronto M5G 1L6, Canada

H. Ngo-Tri
DRF/Biochimie
CEN-G., B.P. 85
38041, Grenoble-Cedex, France

A. Nilsson
Department of Physiological Chemistry
University of Lund
P. O. Box 750
S-220 07
Lund 7, Sweden

Kimiyoshi Ohno
Department of Biochemistry
Sapporo Medical College
West-17, South-1
Sapporo 060 Hokkaido
Japan

R. Paoletti
Institute of Pharmacology and Pharmacognosy
University of Milan
Via A. Del Sarto 21
20129 Milan, Italy

Demetrios Papahadjopoulos
Experimental Pathology
Roswell Park Memorial Institute
666 Elm Street
Buffalo, New York 14203

Haydee E. Pascual de Bazan
Instituto de Investigaciones Bioquimicas
Universidad Nacional del Sur
Bahia Blanca, Argentina

W. A. Pieterson
Laboratory of Biochemistry
State University of Utrecht
Transitorium 3
Padualaan 8, Utrecht, The Netherlands

M. Pilarska
Nencki Institute
3 Pasteur Street
Warsaw, Poland

Robert E. Pitas
Department of Nutritional Sciences
University of Connecticut
Storrs, Connecticut 06268

Giuseppe Porcellati
Istituto de Chimica Biologica
Policlinico
Università di Perugia
Facolta Medica
C.P. n. 3, 06100 Perugia, Italy

Lee Poyer
Biomembrane Research Laboratory
Oklahoma Medical Research Foundation
825 NE 13th Street
Oklahoma City, Oklahoma 73104

Vangala R. Reddy
Worcester Foundation for Experimental
    Biology
Shrewsbury, Massachusetts 01545

N. Shaw
Microbiological Chemistry Research Lab-
    oratories
University of Newcastle-upon-Tyne
Newcastle-upon-Tyne, NE1 7RU England

Patricia Sisson
Agricultural Research Council
Institute of Animal Physiology
Babraham, Cambridge CB2 4AT, England

A. J. Slotboom
Laboratory of Biochemistry
State University of Utrecht
Transitorium 3
Padualaan 8, Utrecht, The Netherlands

C. Spagnuolo
Institute of Pharmacology and Pharmacog-
    nosy
University of Milan
Via A. Del Sarto 21
20129 Milan, Italy

R. Sundler
Department of Physiological Chemistry
University of Lund
P. O. Box 750
S-220 07, Lund 7, Sweden

A. J. Verkley
Laboratory of Biochemistry
State University of Biochemistry
Transitorium 3, Utrecht, The Netherlands

P. H. J. Th. Ververgaert
Biological Ultrastructure Unit
State University of Utrecht
Transitorium 3, Utrecht, The Netherlands

P. M. Vignais
DRF/Biochimie
CEN-G., B. P. 85
38041 Grenoble-Cedex, France

J. J. Volwerk
Laboratory of Biochemistry
State University of Utrecht
Transitorium 3
Padualaan 8, Utrecht, The Netherlands

Moseley Waite
Agricultural Research Council
Institute of Animal Physiology
Babraham, Cambridge CB2 4AT, England

Charles Weddle
Biomembrane Research Laboratory
Oklahoma Medical Research Foundation
825 NE 13th Street
Oklahoma City, Oklahoma 73104

Guey-Shuang Wu
Department of Biological Chemistry
University of California School of Medicine
Los Angeles, California 90024

V. Zambotti
Department of Biological Chemistry
University of Milan
School of Medicine
Via Saldini 50
20133 Milan, Italy

Lipids, Vol. 1, edited by R. Paoletti, G. Porcellati, and G. Jacini. Raven Press, New York © 1976.

Problems Involved with the Purification of Stearoyl-CoA Desaturase

R. Jeffcoat and A. T. James

Unilever Research, Colworth House, Sharnbrook Bedford, England

Although we now understand the general mechanism of desaturation of fatty acids, progress toward the elucidation of the detailed mechanism of these reactions has been slow because of the instability of the enzymes concerned, and because nearly all are firmly bound to some form of cellular membrane structure (1–4). Solubility of the proteins involved is, however, not the complete answer, as was clearly demonstrated by Nagai and Bloch (5) working with the apparently naturally soluble desaturase from *Euglena gracilis*. In spite of the extreme lability of the desaturase fraction with a half-life of less than 10 hr, Nagai and Bloch were, with the aid of standard techniques of protein chemistry, able to resolve the complex into the nonheme iron-sulfur protein ferredoxin, a nicotinamide adenine dinucleotide (NADH) oxidase, and the desaturase enzyme itself. Similar results have been obtained recently by Stumpf with the safflower-seed system.

More recent studies by Holloway (6) and Shimakata et al. (7) working with hen liver and rat liver microsomes, respectively, have confirmed the generality of this tripartite protein system for the desaturase reaction. Using the solubilizing system containing glycerol and sodium deoxycholate, which Lu and Coon (8) had successfully used for the purification of the enzymes of ω-hydroxylation, Holloway was able to resolve the hen liver microsomal $\Delta 9$ desaturase into NADH-cytochrome b_5 reductase, cytochrome b_5, and a cyanide-sensitive protein. Much the same conclusion has been achieved by Shimakata et al. (7) using rat liver microsomes that had been solubilized with 1% (w/v) sodium deoxycholate and 1% Triton X-100, followed by ammonium sulfate fractionation and ion-exchange chromatography on DEAE-Sephadex A-50. In both cases large losses of desaturase activity occurred during the solubilization step. The two most likely explanations for this are (a) denaturation, either reversible or irreversible, of the enzymes by the detergents and (b) removal of a lipid component, which is essential for enzyme activity. Of the two, the simplest to test experimentally is the latter, and indeed there is circumstantial evidence to support this explanation. Wakil and Jones (9) have demonstrated the lipid requirement of the NADH-cytochrome c reductase and have extended their studies to the hen liver stearoyl-coenzyme A (stearoyl-CoA) desaturase system (10). They claim that treatment of hen liver microsomes with 90% aqueous acetone removes all the lipid components from these membranes with a concomitant loss of desaturase activity. This activity can only be restored by the addition of both the phospholipid and neutral lipid fractions. In our laboratory we have been unsuccessful in repeating these experiments, but in attempting to do so we have clearly demonstrated the lack of requirement of the neutral lipid fraction for enzyme activity.

Freeze-dried hen liver microsomes were extracted with 0, 2.5, 5.0, 7.5, and 10% water in acetone. The specific activities of the residual protein fractions were determined, and the results are summarized in Table 1. Analysis of the acetone soluble and insoluble fractions were made using thin-layer chromatography (TLC) on silica gel H with the solvent system ether/petroleum ether/formic acid, 15:85:1. Qualitatively it appeared that 2.5% water in acetone, the concentration that gave the highest specific activity in the residual protein, removed most of the neutral lipid fraction. This was confirmed by quantitative analysis employing preparative TLC and gas-liquid chromatography (GLC). With these techniques it was demonstrated that 2.5% water in acetone removed 97% of the triglyceride, 94% of the free fatty acid, and 99% of the sterols. Estimation of the phospholipid by the method of Rouser et al. (11) before and after extraction with acetone showed that 44% of this fraction had also been removed. For the purposes of enzyme purification, it has now been possible to obtain similar results with wet microsomal pellets extracted with 100% acetone. Extraction of the remaining phospholipid with either acetone or detergents results in irreversible loss of desaturase activity, but whether this is due to loss of essential lipid or denaturation of the enzyme proteins cannot as yet be stated with certainty. Attempts to restore activity by the addition of sonicated preparations of hen liver microsomal phospholipid, free fatty acid, triglyceride, or sterol added separately or together were all unsuccessful. The difficulties involved in the reconstitution of membrane material has been the subject of a recent review by Razin (12).

Removal of nonessential lipid from hen liver microsomes as described above results in preparations that although enzymically active no longer form stable suspensions in aqueous buffer solutions. This makes the problems of solubilization that much more difficult; Gurr and Robinson (13) have reported that, whereas

TABLE 1. *Acetone extraction of lyophilized hen liver microsomes*

Percentage water in acetone	Total units pmoles/min at 30°C	Percentage activity
0	25,163	130
2.5	27,338	142
5.0	17,363	90
7.5	12,450	65
10.0	6,600	34
Control[a]	19,313	100

[a] Control—Lyophilized hen liver microsomes resuspended in 0.1-M potassium phosphate buffer pH 7.4.

1 M potassium phosphate buffer pH 7.4 can solubilize fresh microsomes, acetone-extracted microsomes are not rendered soluble by the same treatment.

The most useful group of compounds used so far for solubilizing membranes are the detergents by virtue of their relatively mild effect on enzymes. They have however been used for the delipidation of membrane proteins, and this can lead to enzyme inactivation through loss of some essential lipid component. Rogers and Strittmatter (14) have successfully delipidated rabbit liver microsomes using 1% (w/v) sodium deoxycholate. Such treatment had no affect on NADH-ferricyanide reductase activity but abolished the NADH-cytochrome b_5 reductase activity, which could only be restored by addition of the lipid.

The neutral detergent Triton X-100 has also been used by Osborne et al. (15) to delipidate bovine rhodopsin. In the two cases cited, the lipid and protein are separated after detergent treatment by gel filtration or ion-exchange chromatography.

Clearly, therefore, if it is suspected that one is dealing with a lipid-requiring enzyme, caution must be exercised when using detergents for solubilization. Either conditions must be arranged that a lipoprotein is isolated or the conditions for reconstitution of the lipoprotein complex must be established. With this in mind, we investigated the affects of various detergents on the activity of hen liver microsomal desaturase. The results of these studies, shown in Fig. 1, indicate that both the nonionic detergents Triton X-100 and Bio-solv and the ionic detergent sodium deoxycholate have severe inhibitory effects on the desaturase activity. However, of the three detergents, Triton X-100 showed the least inhibition at concentrations required to solubilize the enzyme. During the course of this work, Garewal (16) produced a method for the quantitation of Triton X-100, and this enabled us to determine the concentration of the detergent in various preparations and hence relate this activity to the curve shown. The first experiments were designed to assess the importance of protein concentration on solubilization by 1% (w/v) Triton X-100. Two preparations of hen liver microsomes were tested: (a) microsomes that had been stored at $-8°C$ and (b) microsomes which had been lyophilized and extracted with 97.5% acetone. The results, shown in Tables 2 and 3, clearly show two main features. First, solubility of the microsomal protein is strongly concentration-dependent, and, second, removal of part of the microsomal lipid enhances the efficiency of the extraction of the desaturase activity. By a combination of acetone extraction and Triton X-100 solubilization, it has therefore been possible to effect a threefold purification of stearoyl-CoA desaturase in 75 to 100% yield.

It should be emphasized that the conditions described above cannot be used as a general rule for the solubilization of all membrane-bound enzymes. For each system under investigation, the nature and concentration of the detergent used as well as the protein concentration must be uniquely defined. This can best be demonstrated by the following two examples. In Fig. 1 the irreversible inhibition of hen liver desaturases is achieved with concentrations of deoxycholate as low

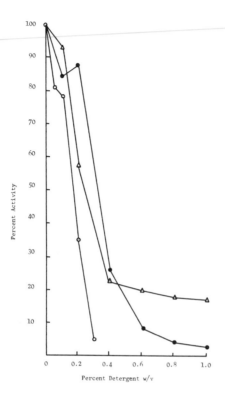

FIG. 1. The stearoyl effect of detergents on the activity of stearyl-CoA desaturase. Incubations contained in a total volume of 1 ml, 100 mmoles potassium phosphate pH 7.4, 100 nmoles NADH, 100 nmoles 1-[^{14}C]stearyl-CoA and varying amounts of sodium deoxycholate (○), Bio-solv. (●), or Triton X-100 (△) as shown.

as 0.3%, whereas rat liver desaturase treated in a similar way still shows between 70 and 75% activity even with concentrations as high as 1% sodium deoxycholate. In a second example (Fig. 2), rat liver and hen liver microsomes are incubated with increasing amounts of Nonidet P-40, a neutral low-molecular-weight detergent used for the isolation of lipoprotein particles from chloroplasts (17). Rat liver desaturase shows a marked stimulation of activity at low detergent concentrations and only 20% inhibition at 1% Nonidet in contrast to the hen liver system, which shows only 20% activity at the higher concentration of detergent. Analysis of the effect of Nonidet on rat liver desaturase indicates that the detergent preferentially solubilizes only part of the desaturase complex, namely the NADH-ferricyanide reductase activity, which is equated with the NADH-cytochrome b_5 reductase activity of liver microsomes (18).

Attempts to solubilize membrane proteins have been made to render them accessible to purification by the standard techniques of protein chemistry. A useful initial step for the purification of water-soluble proteins is that of salt precipitation, but in the presence of bile salts, such as sodium deoxycholate, the range of salt concentration over which proteins are precipitated by ammonium sulfate is reduced to 15 to 35% compared with 15 to 70% saturation in the absence of the bile salts (19). Similarly Triton X-100 also reduces the precipitation

TABLE 2. *Effect of protein/Triton X-100 on solubilization of hen liver microsomes*

Protein mg/ml	Protein solubilized mg	Percentage protein solubilized	Specific activity[a]	Percentage desaturase solubilized
5	2.0	40	927	17.6
10	3.2	32	1,540	23.4
15	3.7	25	1,653	19.4
20	4.9	24	1,320	15.4
25	5.5	22	1,010	10.6
30	5.7	19	467	4.2
Control	—	—	2,107	—

[a] Specific activity is expressed as pmoles·min^{-1}·mg^{-1} and the Triton X-100 concentration was fixed at 1% (w/v).

TABLE 3. *Effect of protein/Triton X-100 on solubilization of partially delipidated hen liver microsomes*

Protein mg/ml	Protein solubilized (mg)	Percentage protein solubilized	Specific activity[a]	Percentage desaturase solubilized
5	2.2	44	627	27.4
10	3.6	36	993	35.5
15	5.4	36	1,120	40.0
20	6.3	32	1,893	59.2
25	7.9	32	2,400	75.3
30	6.0	29	1,800	51.1
Control	—	—	1,007	—

[a] Specific activity is expressed as pmoles·min^{-1}·mg^{-1} and the Triton X-100 concentration was fixed at 1% (w/v).

range and, like the ionic detergents, is itself precipitated by ammonium sulfate. This has the disadvantage of exposing the proteins to high concentrations of detergent, which may or may not have an adverse effect. In the case of the hen liver desaturase solubilized with Triton X-100 in the way described, ammonium sulfate coprecipitated the protein and detergent at 30% saturation of the salt. This treatment had no adverse affect on the enzyme and also failed to effect a purification. This is in contrast to the partial purification of the rat liver desaturase effected by salt precipitation in the presence of Triton X-100 and sodium deoxycholate (7).

Of the standard techniques of protein chemistry, gel filtration and ion-exchange chromatography are perhaps the most commonly used. Gurr and Robinson (13) made use of Sepharose 6B equilibrated with 0.2 M potassium phosphate buffer pH 7.4 to partially purify hen liver desaturase that had been solubilized with 1 M potassium phosphate buffer pH 7.4 according to the method of Scholan and Boyd (20). Gel filtration of the Triton X-100 solubilized material on Sepharose

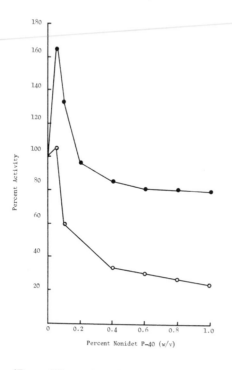

FIG. 2. A comparison of the effect of Nonidet P-40 on rat liver (●) and hen liver (○) microsomal desaturase. Incubation conditions were as for Fig. 1.

6B equilibrated with 0.1 M potassium phosphate buffer pH 7.4 resulted in the elution profile shown in Fig. 3. The first peak eluted with the solvent front contained approximately 26% of the added desaturase, 70% of the eluted reductase, and 20% of the added (and eluted) protein. This represented a 4.5-fold increase in specific activity over the acetone-extracted microsomes, and the protein could be precipitated by centrifugation at $150,000 \times g$ for 2 hr. When a similar column was equilibrated with 1 M potassium phosphate buffer pH 7.4 the elution of the desaturase was only slightly retarded, whereas on Sepharose 4B equilibrated with 0.1 M potassium phosphate buffer pH 7.4 the desaturase was completely retarded. The conclusions from these two columns would therefore suggest that the molecular weight of the aggregate or membrane vesicle to which the desaturase is bound is to the order of 4 to 5×10^6.

The major significance of these findings is that it has been possible to isolate an active soluble preparation in high yield and to demonstrate that on removal of the detergents the lipid to protein ratio is the same as in acetone-extracted microsomes. It is also clear from these results that further purification of the enzyme must be carried out in the presence of detergents, but at the same time care must be taken to ensure that the detergent concentration does not reach such a level that it removes the lipid from what appears to be a lipoprotein complex.

The detailed requirement for lipid is as yet not clearly understood; the only work in this area has been carried out by Rogers and Strittmatter (14). Their

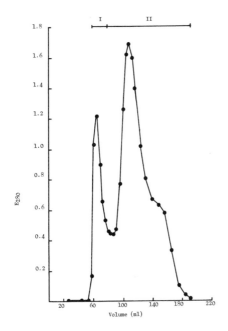

FIG. 3. Elution profile for Triton X-100 solubilized hen liver microsomes on a 20 X 3 cm Sephanose 6B column equilibrated with 0.1-M potassium phosphate pH 7.4.

delipidation of rabbit liver microsomes by sodium deoxycholate solubilization and gel filtration suggested that the function of the lipid in the electron transport chain was one of binding the proteins (cytochrome b_5 and its reductase) together rather than a direct role in the catalytic mechanism. This was based on the observation that NADH-ferricyanide reductase activity was unaffected by removal of the lipid, whereas NADH-cytochrome b_5 reductase was drastically reduced and only restored when sonicated preparations were added in the presence of low concentrations of sodium deoxycholate.

Some indication of the nature of this binding has come from the work of Okuda et al. (21), who have demonstrated a drastic decrease in maximum velocity and a corresponding increase in K_m for NADH-cytochrome b_5 reductase when either protein had been isolated by an enzymic digestion procedure. Spatz and Strittmatter (22, 23) have demonstrated that under these conditions a hydrophobic terminal peptide is cleaved from both cytochrome b_5 and its reductase. Furthermore the work of Takesue and Omura (24) has shown that rat liver cytochrome b_5 reductase isolated by lysosomal digestion is not associated with phospholipid. This is of interest since enzymically solubilized proteins can be purified by conventional methods, and it may therefore suggest that the function of the peptides removed by limited proteolysis is one of binding the lipid required for activity. This has been investigated by Sullivan and Holloway (25), who by sucrose-gradient centrifugation have shown that detergent-isolated but not trypsin-solubilized rabbit liver cytochrome b_5 binds phosphatidylcholine, thereby suggesting a possible role for the nonpolar peptide in this molecule. Similar studies have been carried out by Strittmatter et al. (26) on the binding of cytochrome b_5 to microsomal membranes.

These observations on the general binding properties of microsomal proteins involved in the desaturation of stearoyl-CoA underline some of the problems facing those attempting the purification of such proteins. Not only are there problems of lipid binding that hamper the purification by standard methods, but Okuda et al. (21) have demonstrated that detergent-isolated cytochrome b_5 reductase forms aggregates and is excluded from Sepharose 6B (molecular weight 4 $\times 10^6$). We have shown that even in the presence of Triton X-100 the same enzyme is excluded from G-200 (molecular weight 800,000).

In conclusion it is perhaps worth mentioning the analytic techniques that will be of most use in assessing the criterion of purity of the lipoprotein complex. Considering the problems of aggregation mentioned above and the observations of Okuda et al. (21), who demonstrated that detergent-isolated but not trypsin-isolated cytochrome b_5 can form aggregates with its reductase that are excluded from Sepharose 6B, raises some doubts as to the reliability of polyacrylamide gel electrophoresis in the presence of Triton X-100 as an analytical tool for testing homogeneity. It was therefore decided that purity is best judged by polyacrylamide gel electrophoresis in the presence of sodium dodecylsulfate and β-mercaptoethanol when the proteins are denatured and exist as random coils (27). Analysis of the desaturase eluted from Sepharose 6B gave only three bands on 4% polyacrylamide in the presence of 1% Triton X-100, but the same protein run under the conditions of Weber and Osborn (28) showed good resolution of 10 to 12 distinct electrophoretic regions.

REFERENCES

1. Bloomfield, D. K., and Bloch, K. (1960): *J. Biol. Chem.,* 235:337.
2. Marsh, J. B., and James, A. T. (1962): *Biochim. Biophys. Acta,* 60:320.
3. Harris, R. V., and James, A. T. (1965): *Biochim. Biophys. Acta,* 106:456.
4. Oshino, N., and Sato, R. (1972): *Arch. Biochem. Biophys.,* 149:369.
5. Nagai, J., and Bloch, K. (1966): *J. Biol. Chem.,* 241:1925.
6. Holloway, P. W. (1971): *Biochemistry,* 10:1556.
7. Shimakata, T., Mihara, K., and Sato, R. (1972): *J. Biochem.,* 72:1163.
8. Lu, A. Y. H., and Coon, M. J. (1968): *J. Biol. Chem.,* 243:1331.
9. Jones, P. D., and Wakil, S. J. (1967): *J. Biol. Chem.,* 242:5267.
10. Jones, P. D., Holloway, P. W., Peluffo, O., and Wakil, S. J. (1969): *J. Biol. Chem.,* 244:744.
11. Rouser, G., Siakotos, A. N., and Fleicher, S. (1966): *Lipids,* 1:85.
12. Razin, S. (1972): *Biochim. Biophys. Acta,* 265:241.
13. Gurr, M. I., and Robinson, M. P. (1970): *Eur. J. Biochem.,* 15:335.
14. Rogers, M. J., and Strittmatter, P. (1973): *J. Biol. Chem.,* 248:800.
15. Osborne, H. B., Sardet, C., and Helenius, A. (1974): *Eur. J. Biochem.,* 44:383.
16. Garewal, H. S. (1973): *Anal. Biochem.,* 54:319.
17. Shibuya, I., Honda, H., and Maruo, B. (1968): *J. Biochem.,* 64:571.
18. Takesue, S., and Omura, T. (1970): *J. Biochem.,* 67:267.
19. Penefsky, H. A., and Tzagoloff, A. (1971): *Meth. Enzymol.,* 22:204.
20. Scholan, N. A., and Boyd, G. S. (1968): *Biochem. J.,* 108:27P.
21. Okuda, T., Mihara, and Sato, R. (1972): *J. Biochem.,* 72:987.

22. Spatz, L., and Strittmatter, P. (1971): *Proc. Natl. Acad. Sci.,* 68:1042.
23. Spatz, L., and Strittmatter, P. (1973): *J. Biol. Chem.,* 248:793.
24. Takesue, S., and Omura, T. (1968): *Biochem. Biophys. Res. Commun.,* 30:723.
25. Sullivan, H. R., and Holloway, P. W. (1973): *Biochem. Biophys. Res. Commun.,* 54:808.
26. Strittmatter, P., Rogers, M. J., and Spatz, L. (1972): *J. Biol. Chem.,* 247:7188.
27. Reynolds, J. A., and Tanford, C. (1970): *J. Biol. Chem.,* 245:5161.
28. Weber, K., and Osborn, M. (1969): *J. Biol. Chem.,* 244:4406.

Lipids, Vol. 1, edited by R. Paoletti, G. Porcellati,
and G. Jacini. Raven Press, New York © 1976.

Studies on Bacterial Lipids

Howard Goldfine, G. K. Khuller, and Donald R. Lueking

Department of Microbiology, School of Medicine, University of Pennsylvania, Philadelphia,
Pennsylvania 19174

Progress in the study of the membrane lipids of anaerobic micro-organisms has been slower than with the more widely studied aerobic and facultative organisms (1). In the last few years, however, it has become apparent that the complex lipids of anaerobic bacteria differ significantly from those of other bacteria. Almost every one of the anaerobes examined has been found to contain plasmalogens among its phospholipids, and in many cases the ratio of aldehyde released on acid hydrolysis to lipid phosphorus is greater than 0.5 suggesting that over 50% of the phospholipids contain an alk-1-enyl ether, presumably on C-1 of the sn-glycerol-3-phosphate backbone (2–8). Among six species of clostridia, five have greater than 40% plasmalogens among their glycerophospholipids (2, 5). Rumen bacteria such as *Bacteroides succinogenes* (3), *Peptostreptococcus (Megasphaera) elsdenii* (2, 8), *Ruminococcus flavefaciens* (4), and *Selenomonas ruminantium* are also rich in these alk-1-enyl ether lipids. Plasmalogens have also been found in two species of *Propionibacterium,* in *Veillonella gazogenes* (2), and in *Treponema pallidum* (Reiter) (7).

LIPID COMPOSITION

In bacteria the alk-1-enyl ether chains have been shown to vary in structure from species to species, with straight-chain saturated, monounsaturated, branched, and cyclopropane chains as the predominant types (1). The water-soluble compounds linked to the phosphate of C-3 of glycerol also vary considerably. For example N-methylethanolamine, ethanolamine, serine, and glycerol have been found in bacterial plasmalogens, but choline, which is often present in animal plasmalogens, is rare in bacteria. In only a few cases have fairly complete analyses been done on the phospholipids of anaerobic bacteria. Table 1 presents data on the phospholipid composition of three organisms. It can be seen that each has a distinctly different spectrum of phospholipids. In both *Megasphaera elsdenii,* which was previously called *Peptostreptococcus elsdenii,* and in *Sphaerophorus ridiculosis,* the major phospholipids are ethanolamine phosphoglycerides. In *Clostridium butyricum,* the N-methylethanolamine phosphoglycerides are a major class. *Megasphaera elsdenii,* which has no phosphatidylglycerol or cardiolipin, interestingly has a large amount of a different type of negatively charged

TABLE 1. *Phospholipid composition of anaerobic bacteria*

	C. butyricum[a] (mole % lipid P)	M. elsdenii[b] (mole % lipid P)	S. ridiculosis[c] (mole % lipid P)
Ethanolamine phosphoglycerides	14 (55)	64 (89)	67.5 (26)
N-methylethanolamine phosphoglycerides	38 (78)		
Serine phosphoglycerides	trace	36 (72)	
Glycerol phosphoglycerides	26 (38)[d]		11 (7.8)
Cardiolipin	trace		12 (5.2)

Numbers in parentheses represent percent of phospholipid as plasmalogen.
[a] Data from ref. 5.
[b] Data from ref. 8.
[c] Data from ref. 6.
[d] Data from ref. 12.

phospholipid class, the serine phosphoglycerides. The proportion of each lipid class present as plasmalogen is given in Table 1 in parentheses. In general these vary considerably; some are over 70% in the plasmalogen form. Plasmalogens have not been found in aerobic or even in facultative bacteria, whether the latter are grown aerobically or anaerobically (1). It is also of interest that they are generally not found in plants or fungi, but are found in a wide variety of invertebrate and vertebrate animals. In mammals plasmalogens represent a high proportion of the total lipids in certain tissues, for example brain, heart, skeletal muscle, and kidney. I shall return to the problem of the large evolutionary gap between the two groups of organisms that contain plasmalogens in a discussion of the biosynthesis of these lipids.

FATTY ACID COMPOSITION

In view of their quantitative importance in anaerobic bacteria, we have recently turned our attention to the role of plasmalogens in the bacterial cell membrane. The cell membrane, approximately 70 Å thick, is thought to consist of a fluid mosaic of proteins and lipids, in which the lipids are predominantly arranged in a bilayer (9). Many workers have contributed evidence leading to the concept that the bulk of the lipids must be in a liquid-expanded form, in which the hydrocarbon chains are free to move and the lipid molecules free to diffuse, in order for such membrane-mediated processes as transport, growth, respiration (10, 11), and cell wall synthesis to take place at sufficiently rapid rates. We are seeking to determine whether the plasmalogens of anaerobic bacteria contribute to the fluid properties of the membrane. We grew *C. butyricum* at 37, 30, and 25°C and have analyzed the lipids. At the lower temperature, the proportion of glycerol phosphoglycerides increased by about 40%, whereas the ethanolamine plus N-methylethanolamine phosphoglycerides decreased by 17% (12). This increase in anionic lipids at 25°C may impart a more fluid character to the membranes. According to the work of Papahadjopoulus et al. (13), phosphatidylglycerol vesicles have a lower melting point than phosphatidylethanolamine bilayers with the same fatty acid composition. We have also examined the lipid classes for changes in the proportions of plasmalogens. We found a pronounced increase in the proportion of the glycerolphosphoglycerides present as plasmalogens, from 38% at 37°C to 60% at 25°C.

A wealth of evidence accumulated over the past several years, especially in work with bacterial lipids, has pointed to the central role of unsaturated fatty acids in the maintenance of membrane fluidity. Cyclopropane and terminally branched fatty acids appear to play similar roles when they are present. The ability of bacteria to adjust their acyl chain composition in response to growth temperature has been demonstrated in a number of species. For example the proportion of unsaturated plus cyclopropane fatty acids in *Escherichia coli* increases as the growth temperature is lowered (14, 15). Strain CR34 is typical of *E. coli* strains in this regard, and the ratio of unsaturated plus cyclopropane fatty acids to saturated fatty acids goes from 1.2 at 37°C to 1.4 at 25°C (16). We have looked for similar adjustments in the hydrocarbon chains of *C. butyricum* lipids. As shown in Fig. 1, although the acyl chains of the phospholipids did show a progressive increase in the degree of unsaturation at lower growth temperatures, the alk-1-enyl chains were significantly more saturated at 30°C than at 37°C. When we take into account the fact that there are about two times as many acyl chains as alk-1-enyl chains in the total phospholipids, we see by the dashed line that the total hydrocarbon chains become more unsaturated at lower growth temperatures. Upon examining the individual lipid species, we did not observe increased unsaturation at lower growth temperatures in the ethanolamine plus N-methylethanolamine diacylphosphatides. In fact the fatty acids of these lipids became slightly more saturated at lower growth temperatures. It is of interest to note that the fatty acids at C-1 are more unsaturated than those at C-2, as was first shown by Hildebrand and Law (17). This pattern is unusual in bacterial phospholipids, and we have found that it is maintained over the temperature range from 37°C down to 25°C (12).

The reciprocal changes in fatty acids and alk-1-enyl chains seen in the total

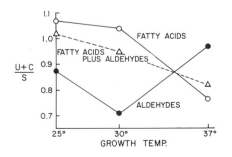

FIG. 1. Ratio of unsaturated (U) plus cyclopropane (C) to saturated (S) fatty acids and alk-1-enyl chains isolated from *C. butyricum* phospholipids in relation to growth temperature.

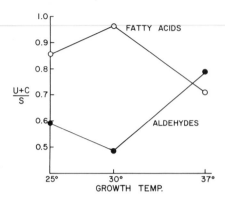

FIG. 2. Ratio of unsaturated (U) plus cyclopropane (C) to saturated (S) fatty acids and alk-1-enyl chains isolated from the ethanolamine and N-methylethanolamine plasmalogens of *C. butyricum* in relation to growth temperature. (From Khuller and Goldfine, 12.)

phospholipids were found in the purified ethanolamine plus N-methylethanolamine plasmalogens (Fig. 2). These are two of the major phospholipids in this species, constituting 30% of the total phospholipids and more than half of the plasmalogens at 37°C. Thus we see that the major plasmalogens and total phospholipids of this organism responded to changes in growth temperature in an unexpected way. The acyl groups at C-2 of the plasmalogens become more unsaturated, at lower growth temperatures, but the alk-1-enyl chains became slightly less so.

We have also been able to modify the acyl and alk-1-enyl chain compositions in another, more dramatic, way. *C. butyricum* is a biotin auxotroph that can be grown in the absence of biotin if fatty acids are added to the medium, as shown by Broquist and Snell (18). Figure 3 shows the changes in the acyl and alk-1-enyl chain composition when cells are subcultured several times at 37°C in media with oleic acid in place of biotin.[1] After three subcultures on oleate, the total 18 : 1 plus 19 : cyclopropane fatty acid (cyc)[2] in the phospholipid acyl chains increased to about 30%. On the other hand, the total 18 : 1 plus 19 : cyc in the alk-1-enyl chains was increased to over 80% after three subcultures.

The average data obtained for cells subcultured four times with oleate at both 37 and 25°C are given in Table 2. At 37°C the 18 : 1 plus 19 : cyc in the acyl chains was 38%; however, the alk-1-enyl chains were 85% 18 : 1 plus 19 : cyc. The total degrees of unsaturation are also given. The acyl chains were not much more unsaturated than those of biotin-grown cells at 37°C, and they were somewhat longer in average chain length. The alk-1-enyl chains were substantially more unsaturated than in biotin-grown cells. When cells were grown with oleate at 25°C, there was a much greater change in the acyl chain composition, with

---

[1] The Casamino acids used in these experiments contained traces of biotin (Nutritional Biochemicals, vitamin-free casein hydrolysate). More recent experiments with vitamin-free casein hydrolysate from the same company especially prepared for microbiologic assays have produced more complete replacement of the acyl chains by 18 : 1 and 19 : cyc in cells grown with oleate.

[2] Unsaturated fatty acids are designated number of carbon atoms : number of double bonds.

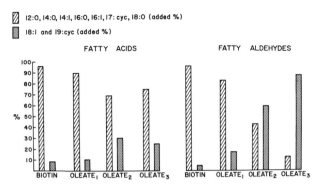

FIG. 3. Changes in the acyl and alk-1-enyl chain composition of *C. butyricum* phospholipids when cells are subcultured several times at 37°C in media with oleic acid in place of biotin. The subscript next to oleate refers to the number of subcultures.

78% of 18 : 1 plus 19 : cyc and 87% total unsaturated. There was little further change in the alk-1-enyl chain composition.

We have also supplemented the growth medium with other fatty acids. The results of linoleate substitution at 37°C are given in Table 3. Again we see a much greater substitution of the alk-1-enyl chains by this diunsaturated fatty acid than there is in the acyl chains. The remaining endogenous fatty acid synthesis is largely reserved for the acyl chains. We have also grown cells with the *trans*-9-18 : 1 acid, elaidic acid. It is known that *trans* fatty acids are between *cis* unsaturated and saturated fatty acids with regard to their melting points and their ability to maintain the liquid-expanded state of membranes at lower temperatures (11). There was a very substantial replacement of both the acyl and alk-1-enyl chains at 37°C, but 24% saturated fatty acids were still found in the acyl chains (Table 3).

TABLE 2. *Phospholipid acyl and alk-1-enyl composition of* C. butyricum *grown on oleic acid*

	Acyl chains (wt %)		Alk-1-enyl chains (wt %)	
	37°C (4)	25°C (2)	37°C (4)	25°C (2)
14:0	7.3	2.2	—	—
16:0	34.0	10.6	3.7	6.2
16:1	13.1	6.9	5.8	1.0
17:cyc	6.7	1.5	5.6	2.7
18:1	22.8	53.3	12.0	21.6
19:cyc	15.1	24.7	72.8	68.6
Saturated	42.0	13.5	3.8	6.2
Unsaturated + cyclopropane	58.0	86.6	96.2	93.9

Numbers in parentheses are number of cultures analyzed.

TABLE 3. *Phospholipid acyl and alk-1-enyl composition of* C. butyricum
grown on linoleic and elaidic acids

	Linoleic acid (wt %)		Elaidic acid (wt %)	
	Acyl chains	Alk-1-enyl chains	Acyl chains	Alk-1-enyl chains
14:0	10.8	—	7.1	—
16:0	40.1	6.1	15.2	4.2
16:1	11.6	11.8	6.0	5.5
17:cyc	3.8	6.1	3.0	4.1
18:1	4.9	5.4	66.6[a]	84.6
18:2	26.2	63.8	—	—
19:cyc	1.3	6.8	0.6	1.4
Saturated	51.7	6.1	23.6	4.5
Unsaturated				
+ cyclopropane	48.1	93.9	76.5	95.6

[a] 89% *trans* 18:1.

We have begun an examination of the physical properties of the phospholipids of *C. butyricum* grown at different temperatures and with various fatty acids in place of biotin. The spin-lattice relaxation time, T_1, of the protons obtained by Fourier transform nuclear magnetic resonance provides information on the motions of the hydrocarbon chains (19, 20). According to theory and the results obtained by other workers, T_1 increases with increasing motion of the chains in the vicinity of the protons under investigation. Our observations of the T_1 of both the methylene and methyl protons of the hydrocarbon chains of vesicles produced by sonicating the total phospholipids of *C. butyricum* in a D_2O buffer, show, as expected, decreased motion at lower temperatures of observation. They also show somewhat greater motion at the methyl ends than in the bulk methylene groups. These measurements do not distinguish between acyl and alk-1-enyl chains and have shown only relatively slight differences between biotin-grown and oleate-grown cells. It appears that changes in the hydrocarbon chains may be compensated for by changes in the proportion of the different lipid classes, which we have observed. Thus the cells appear to be capable of maintaining a relatively constant membrane fluidity under various growth conditions.

PLASMALOGEN BIOSYNTHESIS

As noted above there is an enormous evolutionary gap between these anaerobic bacteria and the invertebrate and vertebrate animals, which also have large amounts of plasmalogens among their phospholipids. The current concepts of plasmalogen biosynthesis in animals has recently been summarized by Snyder (21). The key reactions are an acylation of dihydroxyacetone-phosphate, which is followed by the replacement of the acyl group with a long-chain alcohol, giving rise to a saturated glycerol ether phospholipid. After reduction of the keto group at C-2 and an acylation at that site, a glycerol ether analogue of phosphatidic acid is formed, which after dephosphorylation can undergo reaction with either cytidine diphosphate choline (CDP-choline) or CDP-ethanolamine. The final step in the formation of a plasmalogen in these animal tissues is an oxygen-dependent dehydrogenation of the ether bond to give an alk-1-enyl ether. This step involves cytochrome b_5 and a reduced pyridine nucleotide (22–24).

There are several lines of evidence that argue against this pathway to plasmalogens in anaerobic bacteria. (a) In animals both long-chain alcohols and unsubstituted glycerol ethers are readily incorporated into plasmalogens (21). They are not so used in *C. butyricum* (1, 25). (b) Although many anaerobic bacteria have very small amounts of saturated glycerol ethers, no correspondence has been found between the alcohol and alk-1-enyl chain compositions, as is often found in animal tissues (1). (c) As I have just noted, the formation of the alk-1-enyl bond in animal tissues requires molecular oxygen, which is not available to the strict anaerobes. (d) *In vitro* studies, in my laboratory and in that of Dr. Hagen, have been unable to demonstrate the utilization of either glyceraldehyde 3-P or dihydroxyacetone-P in extracts of *C. butyricum*. (e) Hill and Lands have shown that [1-^{14}C-2-^{3}H]glycerol is readily used for both diacylphosphatide and plasmalogen formation by whole cells of *C. butyricum,* and the tritium is retained in both classes of phospholipids (26). This experiment also eliminates dihydroxyacetone-P, at least from the pathway taken by external glycerol into the plasmalogens.

There are several similarities between bacteria and animals with respect to ether lipid synthesis. First, both types of organisms appear to use fatty acids as sources of the alk-1-enyl chains. We have demonstrated the utilization of long-chain fatty acids as alk-1-enyl chain precursors *in vivo* in *C. butyricum* and the enzymic reduction of long-chain acyl-CoA derivatives to yield both long-chain aldehydes and alcohols (5, 27). We have also shown that exogenous [1-^{14}C-1-^{3}H]palmitaldehyde equilibrates with the fatty acid pool in this organism, but a significant portion is incorporated into the alk-1-enyl chains without loss of tritium (25). Thus incorporation occurs either at the aldehyde level or at a more reduced level, as it does in animal tissues. The question of how the alk-1-enyl ether bond is formed in these bacteria still eludes us.

In the biosynthesis of unsaturated fatty acids, higher organisms desaturate long-chain acyl-CoA esters, a reaction that requires molecular oxygen (28). In anaerobes and in other bacteria, however, unsaturated fatty acids are formed by an anaerobic chain-elongation process, in which the double bond is inserted at the 10-carbon or 12-carbon stage of chain elongation by dehydration of a β-hydroxy acid derivative (28). This distinction between anaerobic and aerobic biosynthetic mechanisms is found in other pathways (29). Plasmalogen biosynthesis appears to be another example of this disunity in biochemistry, in which anaerobic mechanisms exist in bacteria and aerobic mechanisms appear as higher organisms evolve. It is tempting to speculate that plasmalogen synthesis evolved separately in higher organisms in response to a need for specialized lipid structures in highly differentiated nervous and muscle tissues.

PHOSPHATIDIC ACID SYNTHESIS IN BACTERIA

Figure 4 outlines lipid synthesis in *Escherichia coli*. Fatty acids are produced endogenously by a sequence of reactions beginning with the carboxylation of acetyl-CoA to yield malonyl-CoA. All of the subsequent steps of chain elongation involve a small protein with a 4'-phosphopantetheine prosthetic group, the acyl-carrier protein (ACP) (30). As indicated transacylation of the endogenously formed fatty acids to glycero-3-phosphate (G3P) probably occurs directly from ACP without the intermediate formation of acyl-CoA derivatives (31, 32). *E. coli* membrane particles can also transfer acyl groups from acyl-CoA to G3P (31, 33). Both reactions occur *in vitro* at approximately equal rates, which are commensurate with the rate of lipid synthesis in the cell. In *C. butyricum* the preferred acyl donors are acyl derivatives of the ACP, the rate of transacylation to glycerophosphate being 5 to 10 times greater with ACP derivatives than with CoA derivatives (34, 35).

Another distinction between the two organisms is the fatty acid specificity of the G3P acyltransferases. In the reaction catalyzed by glycerophosphate acyltransferase(s) from *E. coli*, saturated fatty acids are strongly preferred (32). On the other hand, the rate of transacylation to G3P was four- to eightfold greater with oleyl-ACP than with palmityl-ACP when membrane particles from *C. butyricum* were tested (35). The results with both organisms are consistent with the distribution of fatty acids on the cellular phospholipids. As noted earlier *C. butyricum* diacylphosphatides are predominantly unsaturated on C-1 and satu-

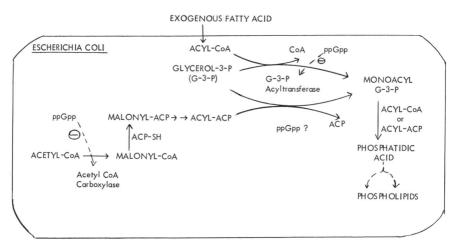

FIG. 4. Lipid synthesis in *E. coli*.

rated on C-2. The more common reverse pattern is found in *E. coli* lipids (17), and the preference for saturated fatty acids for acylation at C-1 is consistent with the presence of largely saturated fatty acids at that position in the phospholipids of this organism.

Both acyl-CoA esters and acyl-ACP esters can be utilized for the second step in the formation of phosphatidic acid, i.e., the acylation of monoacyl G3P catalyzed by *E. coli* membrane particles. van den Bosch and Vagelos (32) have shown that this reaction uses unsaturated fatty acids preferentially with either acyl-ACP or acyl-CoA derivatives, but the reaction is more specific for unsaturated fatty acids when the ACP derivatives are used. In our studies with *C. butyricum,* the ACP derivatives were slightly preferred in the second acylation step as judged by competition experiments, and saturated fatty acids were preferred over unsaturated fatty acids by a ratio of 58:42 (35). Again the results with both *E. coli* and *C. butyricum* indicate that the positioning of saturated and unsaturated fatty acids in bacterial phospholipids may be largely regulated by the acyltransferases to glycerophosphate and monoacylglycerophosphate. In addition Sinensky (36) has demonstrated a temperature effect on this selectivity with acyl-CoA esters, which is consistent with the increased unsaturation of the cellular lipids at lower growth temperatures in *E. coli.*

Phosphatidic acid synthesis also appears to be a site of regulation of the rate of lipid synthesis in bacteria. When a mutant *E. coli* is starved for a required amino acid, there is a strong inhibition of nucleic acid synthesis and the rate of phospholipid synthesis is decreased by two- to fourfold (37). Two compounds accumulate in these starved cells, guanosine tetra- and pentaphosphates (38). Recent work has implicated guanosine tetraphosphate (ppGpp) in the regulation of lipid synthesis in *E. coli.* Polakis et al. (39) have shown a 50% inhibition of acetyl-CoA carboxylase by ppGpp, and Merlie and Pizer (40) have demonstrated a similar inhibition of acyl-CoA : glycerophosphate acyltransferase by the same compound. A similar effect on the synthesis of phosphatidylglycerol *in vitro* was observed (40). We have asked whether ppGpp also inhibits transacylation from acyl-ACP, which we presume to be the donor for endogenously synthesized fatty acids. As did Merlie and Pizer, with the acyl-CoA derivatives we observed a progressive inhibition of glycerophosphate acylation by ppGpp, which produced a 60% inhibition at 4.3 mM ppGpp at a Mg^{2+} concentration of 5 mM (Fig. 5). However, when palmityl-ACP was used as the acyl donor there was no inhibition of glycerophosphate acylation (Fig. 6). The inhibition of palmityl CoA : glycerophosphate acyltransferase was greatest at a Mg^{++} to ppGpp molar ratio of one and our studies on the acylation products show that the acylation of glycerophosphate and not that of monoacylglycerophosphate is specifically inhibited by ppGpp when palmityl CoA is the acyl donor (Table 4).

In summary one site of the regulation of endogenous lipid synthesis by ppGpp appears to be at the level of acetyl-CoA carboxylase as shown by Polakis et al. (39). The inhibition of fatty acid transfer from CoA to glycerophosphate is presumably concerned with the regulation of growth when fatty acids are being

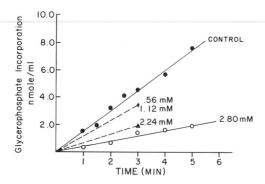

FIG. 5. The effect of ppGpp on the acylation of glycerophosphate by *E. coli* membrane particles with palmityl-CoA as acyl donor. Concentration of ppGpp is shown for each line.

taken up from the medium and utilized for lipid synthesis, a situation that inhibits endogenous fatty acid synthesis in some strains of *E. coli* (41). The recent work of Nunn and Cronan (42) has shown that lipid synthesis is indeed inhibited during amino acid starvation when *E. coli* is dependent on external fatty acids. The differential inhibition of G3P transacylation by ppGpp, depending on the nature of the acyl donor, is an important indication that the two reactions are physiologically distinct and not merely one substrate mimicking the other as a result of a structural similarity. It raises the possibility that there are distinct enzymes or subunits for the two acyl carriers in *E. coli*. This subtle regulatory system points out the importance of controlling phospholipid synthesis in growing cells. Studies of bacterial lipid synthesis should provide additional clues to such regulatory mechanisms, and may eventually be of value in understanding the regulation of lipid synthesis in animal tissues.

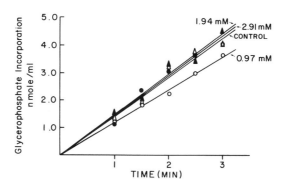

FIG. 6. The effect of ppGpp on the acylation of glycerophosphate by *E. coli* membrane particles with palmityl-ACP as acyl donor. Concentrations of ppGpp are given next to each line.

TABLE 4. *Effect of ppGpp on the distribution of products of the acyltransferase reaction employing palmityl-CoA and palmityl-ACP as acyl donors*

Acyl substrate	Monoacyl-glycerol 3P (nmoles formed)	Phosphatidic acid (nmoles formed)
Palmityl-CoA		
Control	0.527	0.624
4.5 mM ppGpp	0.256	0.714
Palmityl-ACP		
Control	0.104	0.031
4.5 mM ppGpp	0.105	0.044

Reaction mixtures contained 0.60-mM [^{14}C]G3P 5.0-mM mercaptoethanol, 10-mM $MgCl_2$, 100-mM Tris-HCl pH 8.5, 44-μM palmityl-CoA or 63-μM palmityl-ACP, and 0.63 mg/ml of *E. coli* membrane particles. The reactions were stopped after 5 min in reactions with palmityl-CoA and after 3 min in reactions with palmityl-ACP by the addition of an equal volume of 4-N HCl and the lipids were extracted and purified as previously described (35, 36).

ACKNOWLEDGMENTS

This work was supported by U.S. Public Health Service Grant AI-08903 from the National Institute of Allergy and Infectious Diseases. D. R. L. is a recipient of U.S. Public Health Service Postdoctoral Fellowship GM-57521.

REFERENCES

1. Goldfine, H., and Hagen, P. O. (1972): In: *Ether Lipids: Chemistry and Biology*, edited by F. Snyder, p. 329. Academic Press, New York.
2. Kamio, Y., Kanegasaki, S., and Takahashi, H. (1969): *J. Gen. Appl. Microbiol.*, 15:439.
3. Wegner, G. H., and Foster, E. M. (1963): *J. Bacteriol.*, 85:53.
4. Allison, M. J., Bryant, M. P., Katz, I., and Keeney, M. (1962): *J. Bacteriol.*, 83:1084.
5. Baumann, N. A., Hagen, P. O., and Goldfine, H. (1965): *J. Biol. Chem.*, 240:1559.
6. Hagen, P. O. (1974): *J. Bacteriol.*, 119:643.
7. Meyer, H., and Meyer, F. (1971): *Biochim. Biophys. Acta*, 231:93.
8. van Golde, L. M. G., Prins, R. A., Franklin-Klein, W., and Akkermans-Kruyswijk, J. (1973): *Biochim. Biophys. Acta*, 326:314.
9. Singer, S. J. (1971): In: *Structure and Function of Biological Membranes*, edited by L. I. Rothfield, p. 145. Academic Press, New York.
10. Overath, P., Schairer, H. U., and Stoffel, W. (1970): *Proc. Natl. Acad. Sci. USA*, 67:606.
11. Fox, C. F. (1972): In: *Membrane Molecular Biology*, edited by C. F. Fox and A. D. Keith, p. 345. Sinauer Associates, Stamford, Conn.
12. Khuller, G. K., and Goldfine, H. (1974): *J. Lipid Res.*, 15:500.
13. Papahadjopoulos, D., Jacobson, K., Nir, S., and Isac, T. (1973): *Biochim. Biophys. Acta*, 311:330.
14. Marr, A. G., and Ingraham, J. L. (1962): *J. Bacteriol.*, 84:1260.
15. Esfahani, M., Barnes, E. M., Jr., and Wakil, S. J. (1969): *Proc. Natl. Acad. Sci. USA*, 64:1057.
16. Hechemy, K., and Goldfine, H. (1971): *Biochem. Biophys. Res. Commun.* 42:245.
17. Hildebrand, J. C., and Law, J. H. (1964): *Biochemistry*, 3:1304.
18. Broquist, H. P., and Snell, E. E. (1951): *J. Biol. Chem.*, 188:431.

19. McLaughlin, A. C., Podo, F., and Blasie, J. K. (1973): *Biochim. Biophys. Acta,* 330:109.
20. Horwitz, A. F., Klein, M. P., Michaelson, D. M., and Kohler, S. J. (1973): *Ann. N.Y. Acad. Sci.,* 222:468.
21. Snyder, F. (1972): In: *Ether Lipids: Chemistry and Biology,* edited by F. Snyder, p. 121. Academic Press, New York.
22. Paltauf, F. (1972): *Fed. Eur. Biochem. Soc. Lett.,* 20:79.
23. Snyder, F., Wykle, R. L., Blank, M. L., Lumb, R. H., Malone, B., and Piantadosi, C. (1972): *Fed. Proc.,* 31:454.
24. Paltauf, F., Prough, R. A., Masters, B. S. S., and Johnston, J. M. (1974): *J. Biol. Chem.,* 249:2661.
25. Hagen, P. O., and Goldfine, H. (1967): *J. Biol. Chem.,* 242:5700.
26. Hill, E. E., and Lands, W. E. M. (1970): *Biochim. Biophys. Acta,* 202:209.
27. Day, J. I. E., Goldfine, H., and Hagen, P. O. (1970): *Biochim. Biophys. Acta,* 218:179.
28. Bloch, K. (1969): *Acc. Chem. Res.,* 2:193.
29. Goldfine, H., and Bloch, K. (1963): In: *Control Mechanisms in Respiration and Fermentation,* edited by B. Wright, p. 81. Ronald Press, New York.
30. Vagelos, P. R. (1974): In: *Biochemistry of Lipids. Vol. 4, MTP Int. Rev. Science,* p. 99. Butterworths, London.
31. Ailhaud, G. P., and Vagelos, P. R. (1966): *J. Biol. Chem.,* 241:3866.
32. van den Bosch, H., and Vagelos, P. R. (1970): *Biochim. Biophys. Acta,* 218:233.
33. Pieringer, R. A., Bonner, H., Jr., and Kunnes, R. S. (1967): *J. Biol. Chem.,* 242:2719.
34. Goldfine, H., Ailhaud, G. P., and Vagelos, P. R. (1967): *J. Biol. Chem.,* 242:4466.
35. Goldfine, H., and Ailhaud, G. P. (1971): *Biochem. Biophys. Res. Commun.,* 45:1127.
36. Sinensky, M. (1971): *J. Bacteriol.,* 106:449.
37. Golden, N. G., and Powell, G. L. (1972): *J. Biol. Chem.,* 247:6651.
38. Cashel, M., and Kalbacher, B. (1970): *J. Biol. Chem.,* 245:2309.
39. Polakis, S. E., Guchhait, R. B., and Lane, M. D. (1973): *J. Biol. Chem.,* 248:7957.
40. Merlie, J. P., and Pizer, L. I. (1973): *J. Bacteriol.,* 116:355.
41. Silbert, D. F., Cohen, M., and Harder, M. E. (1972): *J. Biol. Chem.,* 247:1699.
42. Nunn, W. D., and Cronan, J. E., Jr. (1974): *J. Biol. Chem.,* 249:3994.

Lipids, Vol. 1, edited by R. Paoletti, G. Porcellati, and G. Jacini. Raven Press, New York © 1976.

Analyses of Subsets of Molecular Species of Glycerophospholipids

A. Kuksis and J. J. Myher

Banting and Best Department of Medical Research, University of Toronto, Toronto, Canada M5G 1L6

Advances in the study and understanding of the metabolism of glycerolipids have been largely dependent upon the progress made in analytical methodology. Application of thin-layer (TLC) (1, 2) and gas chromatography (GC) (3) enabled the extension of the early work of Harris et al. (4), Isozaki et al. (5), and Collins (6) on the metabolic heterogeneity of rat liver phospholipids. The early *in vivo* and *in vitro* work with a variety of metabolic markers has demonstrated that the oligoenoic species may be formed largely via *de novo* pathways while the polyenoic species may be derived via acylation of lysocompounds (see review in ref. 7). Subsequent work has shown that a distinction may be made between biosynthesis of the tetraenes and hexaenes[1] (8, 9), as well as between unsaturation classes containing palmitic and stearic acids (10, 11). Further complications have been introduced by the recognition of the variable nature of the contributions of the minor pathways of glycerophospholipid metabolism to the formation and degradation of specific molecular species (12, 13). Reference may also be made to the recent studies of Kanoh and Ohno (14) and Akesson et al. (15), which suggest that the diacylglycerols derived from phosphatidylcholine (PC) by a reversal of the choline phosphotransferase reaction could contribute significantly to the formation of PC and possibly phosphatidylethanolamine (PE) via the Kennedy pathway (16). Finally differences in the subcellular distribution of the lipid-metabolizing enzymes (7) and in the rates of the exchange of the glycerophospholipids via the exchange proteins (17) would also be expected to contribute to a heterogeneity in the metabolic turnover of the molecular species of glycerolipids.

The recognition of the analytical complexity and metabolic heterogeneity no longer permits the assignment of simple precursor-product relationships to the individual molecular species in the absence of accurate measurements of the pool size and turnover of each precursor, intermediate, and the appropriate final product. Recent work, however, has shown (18) that these goals may be achieved under certain experimental conditions by employing deuterated water as a tracer

---

[1] Oligoenes, trienes, tetraenes, pentaenes, and hexaenes represent molecular species of glycerophospholipids with a total of 0–2, 3, 4, 5, and 6 double bonds per molecule in their diglyceride moiety.

and by analyzing the lipid precursors, intermediates, and final products of the metabolic transformations by GC and mass spectrometry (MS).

THEORY

Studies with tritiated (19, 20) and deuterated (20) water have indicated that the incorporation of isotope into the fatty acids can be used as a reliable measure of the total rate of fatty acid biosynthesis. During the *de novo* synthesis of palmitic acid in 75% D_2O, approximately 50% of the skeletal hydrogen atoms become replaced by deuterium resulting in an accumulation of 10 or more deuterium atoms per molecule. This high degree of labeling is sufficient for both GC and MS resolution of the newly formed and old molecules (18). Recent advances in the understanding of the mechanism of fatty acid biosynthesis allows a rationalization of the deuterium labeling on the basis of rapid and extensive equilibration of the methylene hydrogens of malonate and those of the nicotinamide adenine dinucleotide phosphate (reduced form) (NADPH) reductant with the deuterium in the water. As a result the distribution of deuterium among the newly formed fatty acid molecules becomes largely statistical (18, 20). Therefore heavy labeling by deuterium would be anticipated in both palmitate and stearate, which are the major products of fatty acid synthesis in animal systems, as well as in myristate, which is a minor product. There might be some labeling of the monounsaturated fatty acids derived from any newly formed acids by desaturation, as well as of any old fatty acids subjected to chain elongation. The latter could be recognized by the presence of a limited number of deuterium atoms near the carboxyl end of the chain (20). It would not be possible to distinguish between the stearic acid formed by *de novo* synthesis and that derived by chain elongation from newly formed palmitate. Since water penetrates every compartment, each newly formed acid would become labeled in direct proportion to its rate of synthesis, and its turnover and pool size could be calculated once accurate measurements are obtained of the distribution of these fatty acids among different lipid classes and molecular species. As a result it is possible to obtain a quantitative measure of the percent replacement of the old by the newly synthesized fatty acids in any species of glycerolipid that can be resolved by present methods. Furthermore analysis of the diglyceride (DG) moieties of the glycerophospholipids would reveal combinations of newly formed fatty acids among themselves and with unlabeled fatty acids or fatty acids with a limited deuterium content, which would then be distinguished from combinations of unlabeled fatty acids of the same carbon number. This distribution could be assessed at the level of each molecular species. Glycolysis is by far the major source of glycerol for *de novo* synthesis of glycerolipids. Since the hydrogens of the various 3C phosphates become activated during scission of the glucose-carbon chain and the phosphorylated intermediates themselves undergo reduction, it is obvious that every newly synthesized molecule of glycerol will contain deuterium. An average of three deuterium atoms could be anticipated for each newly formed glycerol molecule at high concentrations of D_2O (19), which ought to be sufficient to provide a satisfactory resolution of old and new molecules of glycerol in the mass spectrometer.

It should be possible to recognize at the monoglyceride level four subsets of molecular species: new glycerol and new fatty acid; new glycerol and old fatty acid; old glycerol and new fatty acid; and old glycerol and old fatty acid. At the DG level, additional subsets would arise from the combination of the above subsets with another new or old fatty acid. Since GC can resolve monoglycerides containing fatty acids with 10 or more deuterium atoms from monoglycerides with less than 10 deuterium atoms in their fatty acid moieties regardless of the extent of labeling of the glycerol moiety, it is possible to recognize combinations of newly formed and old glycerol. The GC resolution of the DGs containing two labeled and two unlabeled fatty acids is also possible, but DGs containing both labeled and unlabeled fatty acids in the same molecule may be more difficult.

The identification and quantitation of the above subsets of molecular species when performed over an appropriate time course should theoretically permit the reassessment of a variety of conclusions about the precursor-product relationships in glycerolipid metabolism.

Specifically the proposed analyses might well provide the much needed differentiation between the utilization of the newly formed and old DGs (DGs generated from hydrolysis of phosphoglycerides or TGs) in the biosynthesis of the various molecular species of glycerolipids. Such analysis would allow the reassessment of the differential utilization of palmitic and stearic acids in glycerolipid metabolism, as postulated by Elovson (21). The determination of the distribution of newly formed palmitate and stearate ought to be free of the criticism of the unnatural nature of the exogenously administered fatty acids. In a comparable manner, the testing could be extended to other established concepts and hypotheses of glycerolipid metabolism, e.g., differential utilization of DGs. Furthermore the new method would allow the examination of new hypotheses about the general nature of glycerolipid metabolism, its regulation, and the location of the various metabolic pathways at the subcellular level. Thus it may be possible to assess the degree of spatial association of the glycolytic pathway and the pathway of fatty acid biosynthesis from the combination frequency of newly formed fatty acids and newly formed glycerol phosphate.

The enthusiastic theoretical considerations presented above must be dampened, however, by caution also justified on theoretical grounds. There is good reason to believe that many of the metabolic transformations involved in the biosynthesis of the fatty acids and glycerol and in their metabolic utilization may be subject to an isotopic effect (22). The deuterated water may equilibrate not only with the hydrogens of those molecules that are involved in lipid metabolism, but also with others, including proteins. As a result the tertiary structure and biochemical properties may be changed. It has been established that well being is impaired in rats in which over 20% of the body water hydrogen has been replaced by deuterium (23). Likewise isolated or cultured cells may not be maintained satisfactorily in buffers containing more than 30% deuterium (24). Other experiments have demonstrated that the presence of elevated amounts of deuterium in a biologic system slows down the biologic clock (25). Therefore it must be recognized that a perfusion of an isolated rat liver with a buffer containing 70 to 90% deuterated water may conjure experimental conditions that are not compatible with normal physiologic function. Nevertheless such experiments appear to yield reliable metabolic data, which have led to the confirmation of several hypotheses based on indirect evidence and have resulted in the validation of experimental procedures. Thus detailed comparisons of the incorporation of deuterium and tritium during fatty acid biosynthesis have given comparable results (20).

EXPERIMENT

In the pioneering work of Rittenberg and Schoenheimer (26), deuterium was supplied in the drinking water to rats, and most of the animals were sacrificed several days later. Recently Curstadt and Sjovall (27) have supplied deuterium in the form of ethanol in the drinking water and have obtained sufficient labeling of lipids for a mass spectrometric analysis by sampling the animals 24 hr later. These methods of administration of the isotope are not suitable for the assessment of the early events in the biosynthesis of the glycerophospholipids from newly formed glycerol and fatty acids due to extensive acyl exchange and recycling. Extensive early labeling of the newly formed fatty acids, however, may be obtained by organ perfusion with concentrated D_2O as shown by Wadke et al. (20). This method was therefore adopted, since it gave effective labeling of the more rapidly formed lipid classes as early as 30 min after the start of the perfusion.

Liver Perfusion

The perfusion of isolated rat livers was performed under standard conditions. Male rats (260 to 300 g) of the Wistar strain were used both as liver and blood donors. The livers were perfused with a red blood cell-buffer system as the perfusion medium, as described elsewhere (28). The concentration of deuterium in the perfusion medium was about 75% after a few passes through the liver. At appropriate periods of time (30 to 300 min), the perfusion was discontinued and samples taken of the liver tissue and perfusate. In several instances the livers were homogenized and the mitochondrial, microsomal, and floating-fat fractions prepared prior to lipid extraction (29).

Lipid Analyses

Total lipid extracts were prepared from the liver tissue, subcellular fractions, and the perfusion medium by the method of Bligh and Dyer (30). Individual neutral lipid (31) and phospholipid (32) classes were separated and recovered by TLC. Molecular species of individual neutral lipid (1) and glycerophospholipid

(2) classes were separated by argentation TLC. The lipid fractions were quantitated by gas-liquid chromatography (GLC) of the fatty acid methyl esters (33), which were prepared by methylation with H_2SO_4-methanol in the presence of silica gel. The positional distribution of the fatty acids in the phosphatide molecules was determined by hydrolysis with phospholipase A_2 (34). Glyceride glycerol was released by hydrogenolysis (35) and was analyzed by GC/MS as described below.

GC/MS Analyses

Combined GC/MS analysis of fatty acid methyl esters (36) and the acetyl esters of monoglycerides (37) were made with a Varian Mat CH-5 single focusing mass spectrometer coupled to a Varian Data 620i Computer as previously described. The GLC separations of the fatty acid methyl esters were made on a Varian Model 2700 Moduline gas chromatograph equipped with a 180 cm $\times$ 2 mm-i.d. stainless steel tube containing 3% SILAR 5CP (a polar methyl cyanopropylsiloxane polymer) on Gas Chrom Q. A similar column was used for the GC/MS determination of deuterated glyceride glycerol as the triacetate. Carrier gas was helium at 10 ml/min. The operating temperatures of the gas chromatograph were 180°C for the fatty acid methyl esters and 270° for the monoglyceride diacetates, with the injector at 225°C and the transfer line at 275°C.

For the determination of deuterium in the methyl palmitate and methyl stearate, the m/e 250 to 350 was scanned repeatedly for the entire GLC run. One run at lower sensitivity was made to check the intensities of the molecular ion M^+ relative to $M^+ + 1$ and $M^+ + 2$. Another run was made at 20 to 50 times the sensitivity and the intensities for the set of heavily labeled molecular ion were measured along with that of the $M^+ + 2$ ion. Deuterated glyceride glycerol was determined from the intensities of the triacetin fragments m/e 103 and 145 caused by loss of CH_3COCH_3 plus CH_3CO and CH_2OCOCH_3, respectively. The determination of deuterium in palmitoylglycerol diacetate was based upon GLC resolution of the species containing newly formed and old palmitate esters. The peak containing the newly formed palmitate, which emerged slightly earlier, was then resolved into molecules containing labeled palmitate and unlabeled glycerol and labeled palmitate and labeled glycerol using MS. Likewise the mass spectrometer was used to resolve the combinations of unlabeled palmitate with unlabeled glycerol, as well as unlabeled palmitate with labeled glycerol. The identification of the labeled glycerol was based on the fragment at m/e 159, which contains all the glycerol hydrogens, one to five of which may be replaced by deuterium atoms. Some correction must be made for the ion at m/e 158. In our experiments this species contained an average of three deuteriums per molecule in the most abundant labeled glycerol moiety. Therefore m/e 159 could be used to monitor species having unlabeled glycerol and m/e 162 (159 + 3) could be used to monitor species having labeled glycerol. The separation of the monoglycerides containing the labeled and unlabeled palmitic acid residues can be recognized from a displacement of the spectrum by about 16 mass units due to an average of 15.8 deuteriums in the newly formed palmitic acid molecules.

Calculations

McCloskey et al. (38) have demonstrated that deuterium-labeled fatty acids are eluted from GLC columns earlier than the corresponding unlabeled molecules. Therefore the fragment ion intensities measured in the mass spectrometer during any single scan required correction for the accumulation of the more heavily labeled molecules in the front and the unlabeled molecules at the tail of the GLC peak. Accordingly, the ion intensities were summed for each mass number over the entire GLC peak coresponding to the appropriate species. Figure 1 illustrates typical labeling patterns observed for methyl palmitate. The labeled molecular ions are found in the region m/e 277 to m/e 294. Old molecules have m/e 270, 271, and 272. The percent relative replacement of a fatty acid (RFA) was calculated as follows:

$$RFA = New\ FA \times 100/New\ FA + Old\ FA$$

The percent relative replacement of a glycerol or monoacylglycerol was calculated by appropriate modifications of the above question. On the basis of the GC/MS findings of the replacement of old by new molecules, we were able to

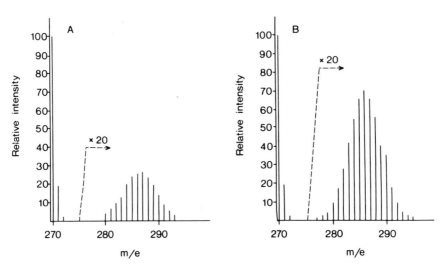

FIG. 1. Typical labeling pattern of palmitic acid during biosynthesis in 75% deuterated water. **(A)** Total lipid. **(B)** free DG.

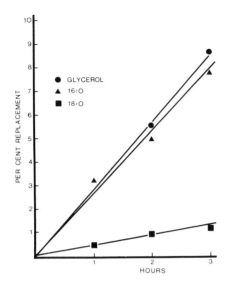

FIG. 2. Rate of hepatic synthesis of total palmitate, stearate, and glyceride glycerol in 75% deuterated water.

calculate the relative turnover rates of the various molecular species of the glycerolipids (39).

RESULTS AND DISCUSSION

Rate of Synthesis of Fatty Acids and Glycerol

Figure 2 gives the rate of synthesis of total palmitic and stearic acids and glyceride glycerol during the perfusion of an isolated rat liver with 75% deuterated water. It is seen that the formation of all the lipid components proceeds linearly over the 3-hr experimental period. Palmitic acid and glycerol appear to be generated at about equal rates, each one reaching about 10% of its total pool at the end of the 3rd hr of perfusion. Stearic acid was formed at about 1/6th the rate of palmitate and glycerol. This rate of synthesis of palmitic and stearic acids is about 25% of that obtained by Wadke et al. (20) who perfused isolated livers of glucose-fed rats, a pretreatment that is known to stimulate fatty acid biosynthesis. Our rate of fatty acid synthesis, however, is close to that observed by Rittenberg and Schoenheimer (26), who allowed chow-fed rats to drink deuterated water and noted that about 50% of total palmitate became labeled in less than 1 day. At our rate of synthesis of palmitic acid (2.5%/hr), it would take about 20 hr to synthesize 50% of the palmitate.

Although effective labeling of glyceride glycerol from tritiated water (19) and from deuterated ethanol (27) has been reported, there have been no previous reports on the quantitative relationships between the formation of new glycerol and new fatty acids.

Rate of Utilization of Newly Formed Fatty Acids in the Biosynthesis of Various Glycerolipid Classes

Figure 3 gives the average rate of replacement of old molecules of palmitic acid by new ones in the various glycerolipid classes of rat liver (28). By far the greatest substitution is seen in the DGs. The rate of replacement appears to be linear in all cases including the DG formation, but the slopes of the lines are different and apparently reflect differential rates of turnover of the various lipid classes, or at least the palmitic acid residues in these lipid classes. The linear rates of replacement of the old molecules of the palmitic acid by new ones in the presumably small and rapidly turning over DG pool is puzzling. Under the conditions of continuous labeling by deuterium, it would have been anticipated that the DGs would have become rapidly and completely labeled. Since this was not the case, it may be concluded that the newly formed DGs were continuously diluted by DGs or fatty acids arising from preferential hydrolysis of the newly synthesized oligoenoic glycerolipids. There was insufficient free fatty acid in the perfusion medium to account for their dilution. It is possible that part of the DG pool originated from a reversal of the cytidine diphosphate choline (CDP-choline) DG

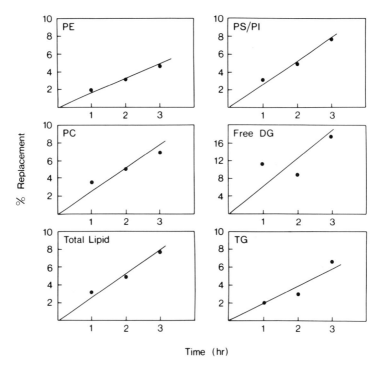

FIG. 3. Average rate of replacement of old molecules of palmitic acid by new ones in various classes of glycerolipids of rat liver.

transferase reaction. This reaction, which has been shown to take place in the brine shrimp (40), could have been a source of both oligoenoic and polyenoic DGs. Linear rates of replacement were also noted for stearic acid in the different fatty acid ester classes, but the low overall incorporation of label in this acid precluded reliable measurements at early times.

Differential Utilization of New Palmitic and Stearic Acids in the Biosynthesis of Different Classes of Glycerophospholipids

Although the rates of utilization of palmitate and stearate were linear in all lipid classes, their slopes showed marked differences. There was evidence that the newly formed stearate was specifically excluded from both DGs and TGs implying that stearic acid was also excluded from phosphatidic acid. Table 1 compares the ratios of utilization of new palmitic and new stearic acid in the biosynthesis of different lipid classes. It is seen that newly formed palmitate is preferred 10-fold over newly formed stearate in the acylation of DGs and TGs when compared to the lecithins, which show a ratio of utilization comparable to that seen in the total. Of the glycerolipids only the mixed fraction of serine and inositol phosphatides showed a ratio of utilization less than that in the total, which suggested a preference for stearate. Clearly the two saturated fatty acids are not equivalent in their metabolic utilization, even when generated *in situ*. Previously Elovson (21) had claimed that exogenous stearate did not become incorporated into DGs *via* the phosphatidic acid pathway, but this conclusion had remained in doubt due to the possibility of unnatural chemical or physico-chemical form of the administered acids.

Table 2 compares the mass distribution of total and newly synthesized palmitate and stearate in the various glycerolipid classes of rat liver at different periods of sampling. Clearly the newly synthesized palmitic and stearic acids were not incorporated into the glycerol esters in a direct proportion to the masses of the existing palmitic and stearic acid esters. Proportionally more of the new palmitate

TABLE 1. *Ratio of utilization of newly formed palmitate and stearate in glycerolipid biosynthesis by rat liver*

Time of perfusion (hr)	Lipid classes (new 16:0 palmitic acid/new 18:0 stearic acid[a])					
	Total	TG	DG	PC	PE	PS + PI[b]
1	5.7	—	74	7.5	5.8	1.9
2	4.8	43	36	7.5	4.3	1.4
3	5.7	51	43	7.1	7.2	2.1

[a] $\dfrac{\text{New 16:0}}{\text{New 18:0}} = \dfrac{\text{R 16:0/100} \times \text{moles 16:0}}{\text{R 18:0/100} \times \text{moles 18:0}}$.

[b] Phosphatidylserine + phosphatidylinositol.

From Kuksis et al., ref. 28.

TABLE 2. *Relative distribution of new palmitate and new stearate in various lipid classes of rat liver*

Time of perfusion (hr)	Lipid classes (moles new/moles total[a])				
	Total	TG	PC	PE	PS + PI
16:0 palmitic acid					
1	1	0.8	1.4	0.8	0.9
2	1	0.8	1.3	0.8	1.3
3	1	1.0	1.1	0.7	1.0
18:0 stearic acid					
1	1		1.1	0.9	0.8
2	1	0.8	0.9	0.8	1.4
3	1	1.2	1.2	0.6	0.9

[a] $\dfrac{\text{Moles new}}{\text{Moles total}} = \dfrac{\text{mole\% new molecules}}{\text{mole\% total molecules}}$.

From Kuksis et al., ref. 28.

was incorporated into the palmitate species than of stearate into the stearate-containing species. These differences were significantly greater at the early times.

Rate of Utilization of Newly Formed Fatty Acids in the Biosynthesis of Molecular Species of Various Glycerolipid Classes

Figure 4 gives preliminary data on the rate of replacement of old molecules of palmitic acid by new ones in the major molecular species of the PC and PE of rat liver (39). In both phosphatide classes, the rates of replacement of palmitate appear to be nonlinear. By the end of the 2nd hr of perfusion, the rate of incorporation of palmitate into the oligoenoic species of both PC and PE starts to level off. This is also true for the hexaenoic PCs, which along with the oligoenes of both phosphatide classes, showed the highest early percent replacement. In contrast the tetraenoic PCs and the hexaenoic PEs appeared to increase their rates of palmitate incorporation with progressing perfusion. The tetraenoic PEs showed a low initial incorporation of deuterated palmitate, just like the tetraenoic PCs, but failed to show subsequent acceleration, resulting instead in a leveling off.

These observations may be attributed to a rapid saturation of a relatively small pool of oligoenoic PCs and PEs, which would give rise to labeled tetraenoic and hexaenoic species of PCs. The hexaenoic species of the latter become more rapidly saturated than the tetraenoic species. This could also be true for the polyenoic PEs, except that in this case the tetraenes would appear to become saturated first, whereas the hexaenes would appear to undergo the acceleration in labeling with progressive perfusion.

This account of the utilization of the newly formed palmitic acid during sustained labeling from deuterated water is consistent with the results of radioactive

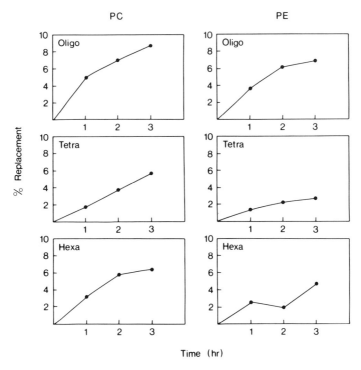

FIG. 4. Rate of replacement of old molecules of palmitic acid by new ones in the major molecular species of PC and PE of rat liver.

pulse labeling of the various molecular species of the choline and ethanolamine phosphatides, which also has suggested a precursor-product relationship between the oligoenoic and polyenoic species (7). The gradually accelerated labeling of some of the polyenes would be due to labeling from both *de novo* synthesis and acyl exchange of the polyunsaturated fatty acids. Since there is as much lower percent of replacement of the stearate, it was not possible to obtain a comparable time course for the percent replacement of stearate in the different molecular species. Only after 3 hr of perfusion did sufficient deuterated stearate accumulate in all species to permit a reliable measurement.

Table 3 compares the ratios of utilization of new palmitate and new stearate in the biosynthesis of the major molecular species of PC and PE. It is seen that the new palmitate is markedly preferred for the biosynthesis of the oligoenes of both phosphatides and of the hexaenoic PEs, in all instances exceeding the ratio in the total about twofold. In contrast the new stearate appears to be utilized preferentially to new palmitate in the biosynthesis of all the other species of both phosphatides, as indicated by a ratio of utilization, which is two to three times below that seen for the total. The relative absence of stearate from the newly formed oligoenes is consistent with their origin via the CDP-nitrogenous base-

TABLE 3. *Molar ratio of utilization of new palmitic and new stearic acid during biosynthesis of molecular species of choline and ethanolamine phosphatides*

Time of perfusion (hr)	Molecular species (new 16:0 palmitic acid/new 18:0 stearic acid[a])					
	Total	Oligoenes	Trienes	Tetraenes	Pentaenes	Hexaenes
PC						
1	7.5	12.4		3.8		5.9
2	7.5	8.4		5.2	1.2	5.9
3	7.1	16.3	6.2	5.5	2.1	8.3
PE						
2	4.3	7.9		2.9		5.2
3	7.3	11.0		3.9		7.7

[a] $\dfrac{\text{New } 16:0}{\text{New } 18:0} = \dfrac{R\ 16:0/100 \times \text{moles } 16:0}{R\ 18:0/100 \times \text{moles } 18:0}$.

From Kuksis et al., ref. 39.

DG phosphotransferase reaction. It was shown above that there was an effective exclusion of newly formed stearate from the DGs and TGs, and presumably from phosphatidic acids. The somewhat lower ratio of new palmitate and new stearate in the oligoenoic phosphatides, compared to the DGs, might have been due to a probable entry of some new stearate by acyl exchange. Acyl exchange would also have to be evoked to account for the presence of relatively much higher amounts of new stearate in the polyenoic species compared to the oligoenes. These observations again are consistent with the findings of radioactive pulse labeling (7), as well as with the distribution of the total mass of fatty acids in these molecular species. In agreement with previous work is also the finding of a high ratio of new palmitate and stearate in the hexaenoic PEs. This species has been shown to become labeled from radioactive glycerol and phosphate to about the same extent as the oligoenoic species, and it has been suggested that it is generated via de novo pathway (9). The high proportion of new palmitate in this species indicates a *de novo* synthesis of the lyso (1-acyl) PE moiety, but the hexaenoic acid must have been introduced by acyl exchange, in view of the absence of hexaenoic species from the DG intermediates. This conclusion requires further verification by studies on the incorporation of labeled docosahexaenoate into this species of PE. The present data would suggest that the hexaenoic PEs do not readily undergo acyl exchange at the *sn*-1-position with newly formed stearic acid.

Differential Utilization of New Palmitate and Stearate in the Biosynthesis of the Molecular Species of Various Glycerophospholipids

Table 4 compares the mass distribution of total and newly synthesized palmitate and stearate in the major molecular species of PC and PE (39). There is

TABLE 4. *Relative distribution of new palmitate and new stearate in major molecular species of choline and ethanolamine phosphatides*

Time of perfusion (hr)	Molecular species (moles new/moles total[a])					
	Total	Oligoenes	Trienes	Tetraenes	Pentaenes	Hexaenes
PC						
16:0 palmitic acid						
1 hour	1	1.4		0.5		0.9
2 hours	1	1.2		0.6	0.8	1.0
3 hours	1	1.2	0.8	0.8	0.8	0.9
18:0 stearic acid						
1 hour	1	1.1		0.9		1.5
2 hours	1	1.2		0.7	1.3	1.4
3 hours	1	1.1	1.0	1.0	0.9	1.1
PE						
16:0 palmitic acid						
2 hours	1	1.8		0.6		1.1
3 hours	1	1.6		0.7		1.1
18:0 stearic acid						
2 hours	1	1.4		0.8		1.3
3 hours	1	1.3		0.9		1.2

[a] $\dfrac{\text{Moles new}}{\text{Moles total}} = \dfrac{\text{mole\% new molecules}}{\text{mole\% total molecules}}$.

From Kuksis et al., ref. 39.

evidence that those species of PC that contain more palmitate (oligoenes) have incorporated more of it than their share. This is consistent with the entry of the palmitate at the time of the DG biosynthesis (41). The distribution of the newly formed stearate is more directly related to the mass of the total stearate in the various molecular species. Since the ratios of new/total stearate in each species are essentially one for most species at all times, it must be concluded that the stearate entry is much more uniform and probably does not involve a precursor-product relationship in the interconversion of molecular species. This could be accounted for if the stearate molecules entered the glycerophospholipids by acyl exchange, possibly as the final step in their metabolic retailoring, and involved all stearate-containing molecular species in a statistical manner.

A distribution related to total mass is also seen in the relative utilization of the newly formed acids in the biosynthesis of the molecular species of the PEs. However, certain significant differences may be pointed out. Since both oligoenes and hexaenes are rich in palmitic acid, they received the bulk of the newly formed palmitate, and both species showed a ratio of new/total palmitate greater than 1.0. The oligoenes received much more than their share of palmitate, which would have been anticipated if the oligoenes were the precursors of the tetraenes. The distribution of the newly formed stearate among the molecular species of the ethanolamine phosphatides was also related to the pre-existing mass of stearate in these species. The major proportion of the total stearate was found in the tetraenes, which also received the bulk of the newly formed stearate, but the ratio of new/total stearate was only 0.9. On the other hand, the ratio of new/total stearate in the oligoenes and hexaenes was 1.2 to 1.3, although these species contained only relatively small proportions of the total stearate. Since the newly formed stearate does not appear to enter the oligoenes readily by *de novo* synthesis, it must have been introduced there preferentially by acyl exchange. The higher proportion of new stearate in the hexaene species cannot be easily attributed to a preferential acyl exchange or *de novo* synthesis as described above. Although the present observations validate several early conclusions about the heterogeneity of metabolism of molecular species of glycerophospholipids and suggest plausible mechanisms, these findings clearly point to the existence of definite subsets of molecular species, which are turned over at markedly different rates. Metabolic studies with the subsets of the molecular species may best be undertaken by determining the time course of association and dissociation of appropriately labeled lipid components in the same molecule. In preliminary experiments we have been able to recognize fragments in the mass spectra of appropriate GLC peaks that are characteristic of the presence of both deuterium-labeled glycerol and deuterium-labeled palmitate (total TG fraction) and deuterium-labeled glycerol and unlabeled palmitate (total PC fraction) in the same molecule of the derived monopalmitin moiety. It was possible that all of the newly formed palmitate had been attached to a newly synthesized glycerol molecule found in the TGs, which further emphasized the *de novo* nature of the entry of palmitate into the glycerolipids. In contrast the finding of a high proportion of newly synthesized glycerol in combination with unlabeled palmitate in the total pool of PC attested to an extensive acyl exchange. This approach promises to provide a much needed measurement of the pool size and turnover rate of the molcular species that participate in each of the transformations, as well as help in the identification of the subcellular site responsible for it. It may then be possible to assess the rate of lipid synthesis as it relates to that of other metabolic activities, such as protein biosynthesis and membrane formation, in the absence of the obscuring effect of conventional measurements of radioactivity and the confusion arising from metabolic reutilization of substrates.

SUMMARY

Theoretical considerations and preliminary metabolic experiments have demonstrated the feasibility of distinguishing between old and new molecules of glycerolipids, resulting in two subsets of each molecular species. Four subsets may be experimentally determined for each molecular species, when analyzed as monoglycerides, by measuring the association of both labeled glycerol and labeled fatty acids with each other and with unlabeled glycerol and fatty acids. Additional subsets of molecular species may be recognized at the DG level by considering the combination of the above four subsets of monoglycerides with newly synthesized and old fatty acids. Still other subsets of molecular species at both monoglyceride and DG level could be recognized on the basis of the association of newly formed and old glycerol and fatty acids among themselves and with fatty acids arising from chain elongation and desaturation. The initial experimental studies were made using isolated rat liver perfused for various times with deuterated water. The lipid analyses were made by GC/MS.

ACKNOWLEDGMENTS

These studies were supported by grants from the Ontario Heart Foundation, Toronto, Ontario, and the Medical Research Council of Canada.

REFERENCES

1. Renkonen, O. (1965): *J.Am. Oil Chem. Soc.,* 42:298.
2. Arvidson, G. A. E. (1965): *J. Lipid Res.,* 6:574.
3. Kuksis, A. (1965): *J. Am Oil Chem. Soc.,* 42:269.
4. Harris, P. M., Robinson, D. S., and Getz, G. (1960): *Nature,* 188:742.
5. Isozaki, M., Yamamoto, A., Amako, T., Sakai, Y., and Okita, H. (1962): *Med. J. Osaka Univ.,* 12:285.
6. Collins, F. D. (1963): *Biochem. J.,* 88:319
7. van den Bosch, H., van Golde, L. M. G., and van Deenen, L. L. M. (1972): *Ergebnisse der Physiology,* p 13. Springer Verlag, New York.
8. Akesson, B., Elovson, J., and Arvidson, G. A. E. (1970): *Biochim. Biophys. Acta,* 210:15.
9. Kanoh, H. (1970): *Biochim. Biophys. Acta,* 218:249.
10. Akesson, B. (1970): *Biochim. Biophys. Acta,* 218:57.
11. Holub, B. J., Breckenridge, W. C., and Kuksis A. (1971): *Lipids,* 6:307.
12. Arvidson, G. A. E. (1968): *Eur. J. Biochem.,* 4:478.
13. Lyman, R. L., Hopkins, S. M., Sheenan, G., and Tinoco, J. (1969): *Biochim. Biophys. Acta,* 176:86.
14. Kanoh, H., and Ohno, K. (1974): In: *12th World Congress of the International Society for Fat Research,* Milan, Italy, p. 22. Abst. No. 12.
15. Akesson, B., Sundler, R., and Nilsson, A. (1974): *12th World Congress of the International Society for Fat Research,* Milan, Italy, p. 23. Abst. No. 13.
16. Kennedy, E. P. (1957): *Ann. Rev. Biochem.,* 26:119.
17. Wirtz, K. W. A., van Golde, L. M. G., and van Deenen, L. L. M. (1970): *Biochim. Biophys. Acta,* 218:176.
18. Kuksis, A., Myher, J. J., Marai, L., Yeung, S. K. F., Steiman, I., and Mookerjea, S. (1974): *Fed. Proc.,* 33:685.
19. Jungas, R. L. (1968): *Biochemistry,* 7:3708.
20. Wadke, M., Brunengraber, H., Lowenstein, J. M., Dolman, J. J., and Arsenault, G. P. (1973): *Biochemistry,* 12:2619.
21. Elovson, J. (1965): *Biochim. Biophys. Acta,* 106:480.
22. Manning, R., and Brindley, D. N. (1972): *Biochem. J.,* 130:1003.
23. Schoenheimer, R., and Rittenberg, D. (1940): *Physiol. Rev.,* 20:218.
24. Orbach, R. (1961): *Proc. R. Soc. Lond.,* A264:458.
25. Enright, J. T., and Vergl, Z. (1971): *Physiology,* 72:1.
26. Rittenberg, D., and Schoenheimer, R. (1937): *J. Biol. Chem.,* 121:235.
27. Curstedt, T., and Sjovall, J. (1974): *Biochim. Biophys. Acta,* 360:24.
28. Kuksis, A., Myher, J. J., Marai, L., Yeung, S. K. F., Steiman, I., and Mookerjea, S. (1975): *Can. J. Biochem.,* 53:509.
29. Rosenfeld, B. (1973): *J. Lipid Res.,* 14:557.
30. Bligh, E. G., and Dyer, W. J. (1959): *Can. J. Biochem. Physiol.,* 3:911.
31. Kuksis, A., Marai, L., Breckenridge, W. C., Gornall, D. A., and Stachnyk, O. (1968): *Can. J. Physiol. Pharmacol.,* 46:511.
32. Skipski, V. P., Peterson, F. R., and Barclay, M. (1964): *Biochem. J.,* 90:374.

33. Kuksis, A. (1971): *Fette Seifen Anstrichmittel,* 73:130.
34. Marai, L., and Kuksis, A. (1967): Lipids, 2:217.
35. Myher, J. J., Marai, L., and Kuksis, A. (1974) *J. Lipid Res.,* 15:586.
36. Myher, J. J., Marai, L., and Kuksis, A. (1974): *Anal. Biochem.,* 62:188.
37. Myher, J. J., and Kuksis, A. (1974): *Lipids,* 9:382.
38. McCloskey, J. A., Lawson, A. M., and Leemans, F. A. J. M. (1967): *Chem. Commun.,* 285.
39. Kuksis, A., Myher, J. J., Marai, L., Yeung, S. K. F., Steiman, I., and Mookerjea, S. (1975): *Can. J. Biochem.,* 53:519.
40. Ewing, R. D., and Finamore, F. J. (1970): *Biochim. Biophys. Acta,* 218:463.
41. Gosselin, L. (1974) In: *12th World Congress of the International Society for Fat Research,* Milan, Italy, p. 23. Abst. No. 14.

Lipids, Vol. 1, edited by R. Paoletti, G. Porcellati, and G. Jacini. Raven Press, New York © 1976.

The Role of 1,2-Diacylglycerol:CDP-Choline (ethanolamine)–Choline(ethanolamine) Phosphotransferase in Forming Phospholipid Species in Rat Liver

Hideo Kanoh and Kimiyoshi Ohno

Department of Biochemistry, Sapporo Medical College, West-17, South-1, Sapporo 060 Hokkaido, Japan

Recently we have studied the metabolic heterogeneity of molecular species of phospholipids in animal tissues. Previously we incubated liver slices with radioactive precursors and found some features of the biosynthetic routes in forming liver phospholipid species (1). The results are summarized in Table 1. More detailed work on metabolic heterogeneity has been reported by other laboratories (2–7). *In vivo* experiments have clearly detected the characteristics of each biosynthetic pathway as summarized in Table 1. However, very little work with cell-free systems has been successful in explaining the selective formation of phospholipid species by various biosynthetic pathways.

The enzymes of *de novo* synthesis, namely choline and ethanolamine phosphotransferase, appeared to form *in vivo* principally dienoic phosphatidylcholine (PC) and hexaenoic phosphatidylethanolamine (PE), respectively, from the common precursor, 1,2-diacylglycerol. The involved reaction is as follows: 1,2-diacylglycerol + cytidine diphosphate (CDP)-choline (ethanolamine) ⟷ PC (PE) + cytidine monophosphate (CMP). The substrate selectivity of the transferases has been studied by incubating liver microsomes with 1,2-diacylglycerol emulsion.

TABLE 1. *Characteristics of biosynthetic routes in forming phospholipid species in rat liver slices*

Pathways	Major products
De novo synthesis	2-Linoleyl-PC
	2-Hexaenoyl-PE
N-methylation of PE	2-Hexaenoyl-PC
Lysophospholipid reacylation	2-Arachidonyl-PC and PE

Summarized from Kanoh, ref. 1.
Rat liver slices were incubated with various radioactive precursors. From the labeling pattern of molecular species of phospholipids, the general features of biosynthetic routes were elucidated.

According to the results reported so far (8–10), both transferases have been shown to possess little specificity, if any, in utilizing various species of 1,2-diacylglycerols. Although we noted some preferential utilization of hexaenoic diacylglycerol by ethanolamine phosphotransferase (10), the observed selectivity could not account for the markedly selective formation of hexaenoic PE as found in liver slices (1) or in *in vivo* experiments (3–5). Apparently the use of the artificial substrates, such as diacylglycerol emulsion, was a major difficulty in studying in detail the properties of the transferases.

BACK-REACTION OF CHOLINE (ETHANOLAMINE) PHOSPHOTRANSFERASE

As shown by Weiss et al. (11), the activity of choline phosphotransferase is reversible. The transferase can degrade PC in the presence of CMP. We prepared liver microsomes labeled in the ionic groups of endogenous phospholipids. By incubating the labeled microsomes with CMP, the formation of CDP-choline or CDP-ethanolamine could be followed. Thus we could study the properties of choline and ethanolamine phosphotransferase through their back-reaction without the use of the substrate emulsion.

The results of this experiment (12) are summarized in Table 2.

The activity of choline phosphotransferase is much higher than that of ethanolamine phosphotransferase, although both transferases have the same affinity for CMP. The product inhibition by CDP-ethanolamine is much stronger than that by CDP-choline. It was also found that choline phosphotransferase has no significant selectivity in degrading endogenous PC species as far as the unsaturated fatty acids esterified in the 2-position are concerned. We suggested that the action of choline phosphotransferase may be degradative as well as biosynthetic *in vivo,* whereas ethanolamine phosphotransferase may operate mainly

TABLE 2. *Degradation of microsomal phospholipids by the back-reaction of choline and ethanolamine phosphotransferase*

	Choline phosphotransferase	Ethanolamine phosphotransferase
Rate	4 to 5 nmoles/mg/min	Less than 0.8 nmole/mg/min
K_m for CMP	0.19 mM	0.14 mM
Product inhibition (K_i)	1.0 mM for CDP-choline	0.05 mM for CDP-ethanolamine
Substrate specificity	Equal utilization of PC species	
In vivo?	Biosynthetic and degradative	Mainly biosynthetic

Summarized from Kanoh and Ohno, ref. 12.
Rats were injected with radioactive choline, ethanolamine, L-methionine, or 1-acyl-lyso-PC, and the labeled microsomes were obtained. The labeled microsomes were incubated with CMP and the formation of CDP-choline or CDP-ethanolamine from endogenous phospholipids was studied.

biosynthetically. Several workers reported data from *in vivo* experiments (13–15) that show the effects of the back-reaction of choline phosphotransferase.

In rat lung we also demonstrated a rapid equilibration between tetraenoic species of diacylglycerol and PC after injection of radioactive palmitic acid, suggesting that in rat lung the back-reaction of choline phosphotransferase operates to supply diacylglycerols from PCs (16). Quantitative and physiologic significance of the back-reaction of choline phosphotransferase in the control of phospholipid metabolism remains to be seen, but the recent results obtained by Sundler et al. in isolated hepatocytes (17) showed that especially the polyunsaturated species of PE may be formed from diacylglycerols derived from PCs by the action of choline phosphotransferase.

MEMBRANE-BOUND 1,2-DIACYLGLYCEROLS DERIVED FROM PCs

To study 1,2-diacylglycerols formed from the microsomal PCs by the action of choline phosphotransferase, we injected rats intraportally with 1-[³H]palmityl-lyso-PC and prepared labeled liver microsomes (18). The labeled microsomes were incubated with CMP; the results of the incubation are presented in Fig. 1. As can be seen, because of the microsomal lipolytic activity, a degradation of PC occurred without the addition of CMP, and diacylglycerols formed in the presence of CMP could not accumulate within the microsomal membranes. We could overcome this difficulty by preincubating the microsomes with di-isopropylfluorophosphate (DFP) as shown in Fig. 2. By treating the microsomes with DFP, the activity of choline phosphotransferase in degrading endogenous PC remained unaffected, whereas the microsomal lipolytic activity was completely suppressed,

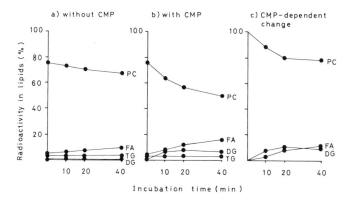

FIG. 1. Time course of incubation of microsomes labeled with 1-[³H]palmityl-lyso-PC. 1-[³H]palmityl-lyso-PC complexed to bovine serum albumin was injected in rats intraportally. After 30 min the liver microsomes were prepared and incubated in the presence (b) or absence (a) of CMP at 37°C; (c) shows the CMP-dependent change. The contents of radioactivity in the lipid components were determined by thin-layer chromatography. FA, free fatty acids; TG, triacylglycerols; and DG, 1,2-diacylglycerols.

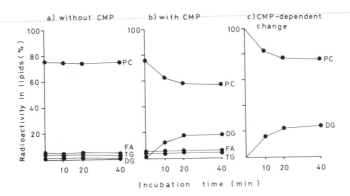

FIG. 2. Time course of incubation of DFP-treated microsomes labeled with 1-[^{3}H]palmityl-lyso-PC. The labeled microsomes were preincubated with 3-mM DFP at 37°C for 1 hr before the experiments. Other details are the same as those given in Fig. 1.

resulting in the accumulation of the formed 1,2-diacylglycerols within the membranes. We then actually determined the content of microsomal 1,2-diacylglycerols after incubation with CMP. As presented in Table 3, the DFP-treated microsomes contained about 90 nmoles of 1,2-diacylglycerols per milligram of protein after a 20-min incubation with CMP. The molecular composition of the formed diacylglycerols was found to be closely related to that of the original microsomal PCs (12), and the major species was of the tetraenoic type.

It was noted that 1,2-diacylglycerols formed by the action of choline phosphotransferase can be much more highly unsaturated than those synthesized from phosphatidic acids *in vivo*, because mainly oligoenoic phosphatidic acids have been reported to be formed in rat liver (3–5, 19). In the next study, DFP-treated

TABLE 3. *Analysis of microsomal 1,2-diacylglycerols*

Experiment	Contents of the endogenous 1,2-diacylglycerols nmoles/mg protein (*n* =6)	Subfractionation of 1,2-diacylglycerols in Experiment B Subfractions	Mole% (*n* =3)
A—The original microsomes	9.7	Mono-	8.1
		Di-	17.5
B—After standard incubation of DFP-treated microsomes	91.6	Tri-	6.9
		Tetra-	45.6
C—After standard incubations of untreated microsomes	39.2	Hexaenoic	22.0

Reproduced from Kanoh and Ohno, ref. 18.
Unlabeled microsomes were incubated with CMP for 20 min with (B) or without (C) preincubation with 3mM DFP. After incubations, microsomal 1,2-diacylglycerols were quantitatively determined. 1,2-Diacylglycerols in DFP-treated microsomes were subfractionated on AgNo$_3$-impregnated thin-layer plates.

nonlabeled microsomes were recovered from the reaction mixture after incubation with CMP for 20 min. The recovered microsomes were reincubated with radioactive CDP-choline or CDP-ethanolamine. We found that the endogenous 1,2-diacylglycerols derived from PCs can be very actively utilized for the formation of PCs and PEs (18). In a given condition, the reaction was linear up to 2 min and the initial rate of formation of the PCs and PEs was 7.4 and 4.5 nmoles/min/mg of microsomal protein, respectively. As presented in Table 3, these endogenous 1,2-diacylglycerols are the mixture of various species with different degrees of unsaturation. We therefore analyzed by argentation thin-layer chromatography the molecular species of phospholipids formed from the endogenous 1,2-diacylglycerols. As the concentration of 1,2-diacylglycerols is very low in the reaction mixture, the results are expressed as a conversion rate of each diacylglycerol species to the corresponding species of phospholipids. The results are given in Fig. 3 (20). It was found that each diacylglycerol species was converted to PC without marked differences. As described in the back-reaction of choline phosphotransferase, the transferase apparently does not possess significant selectivity in either forming or degrading PCs.

In contrast to the action of choline phosphotransferase, ethanolamine phosphotransferase is selective in utilizing hexaenoic diacylglycerol. In the early period of incubation, hexaenoic species comprised about 60% of the newly formed PE. The utilization of diacylglycerols other than hexaenoic type was very low for the fomation of PE, and no appreciable differences were observed in the conversion rate. From the results of the mode of utilization of endogenous substrates by the microsomal enzymes, we believe that the marked synthesis of hexaenoic PE found

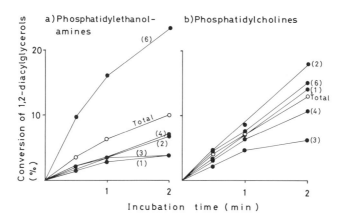

FIG. 3. The conversion rate of the microsomal endogenous 1,2-diacylglycerol species to the corresponding species of the phospholipids. Nonlabeled microsomes pretreated with DFP were first incubated with CMP for 20 min and were recovered by ultracentrifugation from the reaction mixtures. The recovered microsomes were reincubated with radioactive CDP-choline or CDP-ethanolamine and the formed phospholipids were analyzed by argentation thin-layer chromatography. (1), mono-; (2), di-; (3), tri-; (4), tetra-; and (6), hexaenoic species of the phospholipids. (From H. Kanoh and K. Ohno, ref. 20.)

in *in vivo* studies (2–6) should be mainly due to the substrate selectivity of ethanolamine phosphotransferase in utilizing 1,2-diacylglycerol. On the other hand, the products of choline phosphotransferase may be a reflection of the mode of synthesis of the precursor, phosphatidic acid. In this context we should note that there can be a constant supply of hexaenoic diacylglycerol from PCs by the action of choline phosphotransferase, since the product of N-methylation of PEs in rat liver has been shown to be mainly hexaenoic PC (1, 2).

SATURATED FATTY ACIDS LOCATED IN 1-POSITION OF ENDOGENOUS SUBSTRATES

So far we have been concerned with the molecular species of phospholipids with regard to the unsaturated fatty acids esterified to the 2-position of the phospholipids. To study the substrate-selectivity of choline and ethanolamine phosphotransferase in using endogenous substrates with different saturated fatty acids in the 1-position, we injected rats with 1-acyl-lyso-PC labeled with radioactive myristic, palmitic, and stearic acids. Table 4 shows the labeling pattern of the microsomal lipids found after injecting lyso-PC. Most of the incorporated radioactivity was found in the PCs. The radioactivity found in lipids other than PCs did not affect the experiments since the microsomes were treated with DFP (see Fig. 2). Approximately 85 to 95% of radioactivity incorporated into PCs was found to be located in the 1-position. Using argentation as well as reversed-phase partition thin-layer chromatography, we confirmed that over 90% of the fatty acids incorporated into PCs remained unchanged from the original radioactive acids administered in the form of 1-acyl-lyso-PCs.

In any of the three radioactive lysophospholipids, tetraenoic PC was found to be most active in taking up the injected lysophospholipids, in agreement with

TABLE 4. *Distribution of radioactivity in microsomal lipids after injection of labeled 1-acyl-lyso-PC*

	Lyso-PC injected		
Lipid class	1-[^{14}C]myristyl (%)	1-[^{3}H]palmityl (%)	1-[^{3}H]stearyl (%)
Triacylglycerol	4.0	4.1	3.3
Free fatty acids	7.4	4.5	4.5
Diacylglycerols	1.2	0.7	0.9
Phospholipids	87.4	90.7	91.3
	In phospholipids		
PE	5.7	3.4	6.2
PC	84.9	83.3	82.6
Others	9.4	13.3	11.2

From H. Kanoh and K. Ohno, ref. 20.
Radioactive lyso-PC complexed to bovine serum albumin was injected intraportally, and after 30 min liver microsomes were prepared.

our previous results (1, 12, 18). These data show that the intraportally administered 1-acyl-lyso-PC was incorporated into hepatocytes mostly without prior degradation. The labeled microsomes were pretreated with DFP (see Fig. 2) and then incubated with CMP to study the activity of choline phosphotransferase in degrading endogenous PCs labeled differently in position 1. Figure 4 shows the time course of degradation by choline phosphotransferase of microsomal PCs with different labels in the 1-position. The sole product of the reaction was 1,2-diacylglycerol as confirmed by thin-layer chromatographic analysis. As seen in Fig. 4, the 1-myristyl type appeared to be utilized most rapidly, and in decreasing order, 1-palmityl and 1-stearyl species. It is difficult to explain the appreciable difference in activity of the transferase in degrading differently labeled endogenous PCs. Rat liver microsomal PCs have been shown to contain almost equal amounts of palmitic and stearic acids, but the content of the 1-myristyl species has been shown to be very small (21).

The possibility of the presence of different pools of PC species within the membranes seems unlikely, since we found no differences in the activity of choline phosphotransferase in degrading various species of phospholipids synthesized through different biosynthetic pathways (12). We believe that the different reactivity of choline phosphotransferase may be most reasonably explained by assuming that the transferase has a selectivity toward the saturated fatty acids located in the 1-position of the substrates and prefers the fatty acid species of shorter

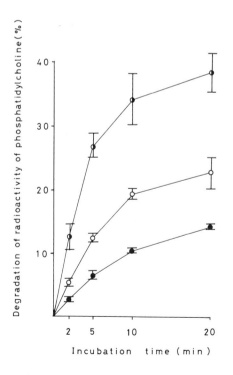

FIG. 4. Degradation by choline phosphotransferase of the endogenous PCs labeled differently in the 1-position. After treatment with DFP, microsomes labeled with 1-[^{14}C]myristyl-, 1-[^{3}H]palmityl- or 1-[^{14}C]stearyl-lyso-PC were incubated with CMP. Average values of three independent experiments are shown with the range of SEM. ●—●, 1-[^{14}C]myristyl; ○—○, 1-[^{3}H]palmityl; and ●—●, 1-[^{14}C]stearyl PC. (From H. Kanoh and K. Ohno, ref. 20.)

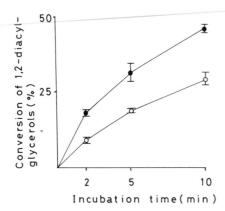

FIG. 5. Utilization by ethanolamine phosphotransferase of 1-[³H]palmityl and 1-[³H]stearyl diacylglycerols. DFP-treated microsomes labeled with 1-[³H]palmityl- or 1-[³H]stearyl-lyso-PC were recovered after incubation with CMP for 20 min. The recovered microsomes were incubated with nonradioactive CDP-ethanolamine. ○—○, 1-[³H]palmityl; and ●—●, 1-[³H]stearyl diacylglycerols. (From H. Kanoh and K. Ohno ref. 20.)

chain length. It is interesting to see that choline phosphotransferase has a selectivity for saturated fatty acids in the 1-position and not for unsaturated fatty acids in the 2-position of the substrates. The physiologic significance of the selectivity of choline phosphotransferase remains to be established, but very recently Sundler et al. (22) observed in isolated hepatocytes that in the *de novo* synthesis of PCs, dimyristyl diacylglycerol was utilized more preferentially than the dipalmityl type and the utilization of distearyl species was the slowest. Their observation may be a reflection of the substrate selectivity of the transferase toward the saturated fatty acids as revealed by our research.

In the next experiments, DFP-treated microsomes labeled with 1-[³H]palmityl and 1-[³H]stearyl-lyso-PC were first incubated with CMP for 20 min. The microsomes were recovered from the reaction mixture by ultracentrifugation and were reincubated with nonradioactive CDP-ethanolamine. By measuring accurately the change in radioactivity of the lipid components, we could follow the utilization of endogenous 1,2-diacylglycerols by ethanolamine phosphotransferase. As shown in Fig. 5, 1-stearyl diacylglycerol was found to be converted to PE relatively more preferentially than the 1-palmityl type. But the selectivity of ethanolamine phosphotransferase (see Fig. 3) for hexaenoic diacylglycerol was much more marked than that for the 1-stearyl type, and these properties of the transferase agree with the results obtained by Sundler in *in vivo* experiments (23).

ENDOGENOUS (MEMBRANE-BOUND) AND EXOGENOUS (EMULSIFIED) SUBSTRATES

From the work thus described, we can study choline phosphotransferase by several procedures, that is, by the use of diacylglycerol emulsion as has been done by many laboratories, through the back-reaction in degrading microsomal PCs, or by utilization of the endogenous 1,2-diacylglycerols formed by the action of the transferase itself.

Since we obtained rather discrepant properties of ethanolamine phosphotransferase from those we previously reported with diacylglycerol emulsion (10), we compared the properties of choline phosphotransferase in utilizing endogenous substrates with those obtained by the use of 1,2-diacylglycerol emulsion. We found that the treatment of microsomes with DFP did not affect appreciably the utilization of endogenous substrates, in either forming or degrading PC. On the other hand, the activity of the microsomal enzyme in utilizing exogenously added diacylglycerol emulsion was completely inhibited by the DFP treatment. In addition we noted that oleate is stimulatory for choline phosphotransferase when using diacylglycerol emulsion in agreement with the results of Sribney and Lyman (24), but rather inhibitory when endogenous substrates were used. McMurray (25) also noted the same inhibitory effects of oleate on the activity of choline phosphotransferase in utilizing endogenous diacylglycerols.

It is difficult to understand why there are differences in the activity of the transferase between the utilization of endogenous and exogenous substrates. These data may at least suggest that the results obtained by using endogenous substrates are not always directly comparable to those obtained by using exogenously added substrate emulsion.

SUMMARY

A physiologic significance of the back-reaction of choline phosphotransferase is suggested, especially in supplying polyunsaturated 1,2-diacylglycerols. We also show a significant selectivity of the microsomal enzymes for *de novo* synthesis of phospholipids in utilizing endogenous substrates with regard to the fatty acids located in 1-position as well as 2-position of the substrates. The discrepant results obtained with endogenous and exogenous substrates are also described.

ACKNOWLEDGMENTS

This work has been supported in part by a scientific research grant from the Ministry of Education, Japan and also from Takeda Chemical Industries, Japan and Nihon Shoji Co. Japan. We thank Miss Asanome, who helped us in preparing the manuscript.

REFERENCES

1. Kanoh, H. (1969): *Biochim. Biophys. Acta*, 176:756.
2. Arvidson, G. A. E. (1968): *Eur. J. Biochem.*, 4:478.
3. Åkesson, B., Elovson, J., and Arvidson, G. A. (1970): *Biochim. Biophys. Acta*, 210:15.
4. Åkesson, B., Elovson, J., and Arvidson, G. A. (1970): *Biochim. Biophys. Acta*, 218:44.
5. Åkesson, B. (1970): *Biochim. Biophys. Acta*, 218:57.
6. Vereyken, J. M., Montfoort, A., and Van Golde, L. M. G. (1972): *Biochim. Biophys. Acta*, 260:70.
7. Holub, B. J., Breckenridge, W. C., and Kuksis, A. (1971): *Lipids*, 6:307.
8. Mudd, J. B., Van Golde, L. M. G., and Van Deenen, L. L. M. (1969): *Biochim. Biophys. Acta*, 176:547.
9. De Kruyff, B., Van Golde, L. M. G., and Van Deenen, L. L. M. (1970): *Biochim. Biophys. Acta*, 210:425.

10. Kanoh, H. (1970): *Biochim. Biophys. Acta,* 218:249.
11. Weiss, S. B., Smith, S. W., and Kennedy, E. P. (1958): *J. Biol. Chem.,* 231:53.
12. Kanoh, H., and Ohno, K. (1973): *Biochim. Biophys. Acta,* 306:203.
13. Bjørnstad, P., and Bremer, J. (1966): *J. Lipid Res.,* 7:38.
14. Bjerve, K. S., and Bremer, J. (1969): *Biochim. Biophys. Acta,* 176:570.
15. Treble, D. H., Frumkin, S., Balint, J. A., and Beeler, D. A. (1970): *Biochim. Biophys. Acta,* 202:163.
16. Moriya, T., and Kanoh, H. (1974): *Tohoku J. Exp. Med.,* 112:241.
17. Sundler, R., Åkesson, B., and Nilson, Å. (1974): *Biochim. Biophys. Acta,* 337:248.
18. Kanoh, H., and Ohno, K. (1973): *Biochim. Biophys. Acta,* 326:17.
19. Hill, E. E., Husbands, D. R., and Lands, W. E. M. (1968): *J. Biol. Chem.,* 243:4440.
20. Kanoh, H., and Ohno, K. (1975): *Biochim. Biophys. Acta,* 380:199.
21. Colbeau, A., Nachbaur, J., and Vignais, P. M. (1971): *Biochim. Biophys. Acta,* 249:462.
22. Sundler, R., Åkesson, B., and Nilsson, Å. (1974): *J. Biol. Chem.,* 249:5102.
23. Sundler, R. (1973): *Biochim. Biophys. Acta,* 306:218.
24. Sribney, M., and Lyman, E. M. (1973): *Can. J. Biochem.,* 51:1479.
25. McMurray, W. C. (1974): *Biochem. Biophys. Res. Commun.,* 58:467.

Lipids, Vol. 1, edited by R. Paoletti, G. Porcellati, and G. Jacini. Raven Press, New York © 1976.

Biosynthesis of Molecular Species of Glycerolipids in Isolated Rat Hepatocytes

B. Åkesson, R. Sundler, and Å. Nilsson

Department of Physiological Chemistry, University of Lund, S-220 07 Lund 7, Sweden

Recent work in our laboratory has aimed to study the relative role of different pathways for the synthesis of different molecular species of glycerolipids in rat liver and to study the specificity against molecular species in different reactions. Studies from other laboratories have indicated that the permeability of liposomes made from different phosphatidylcholine (PC) species is dependent on fatty acid composition (1–3). Furthermore cells grown in the presence of different fatty acids exhibit different functional properties (4). In the light of such observations, it is of interest to elucidate the mechanisms whereby the composition of molecular species in membrane lipids in the tissues are regulated. The liver is of particular interest since, in addition to membrane lipids, lipoprotein and bile lipids are synthetized in this organ.

UTILIZATION OF 1,2-DIACYL-*SN*-GLYCEROLS FOR THE SYNTHESIS OF TRIACYLGLYCEROLS, PCs AND PHOSPHATIDYLETHANOLAMINES

The formation of three dominant classes of glycerolipids from 1,2-diacyl-*sn*-glycerols in the liver was shown by Kennedy (5). When this pathway was studied *in vivo* by injection of [³H]glycerol into rats, the label in liver lipids quickly passed through phosphatidic acids and diacylglycerols and accumulated in triacylglycerols, PCs, and phosphatidylethanolamines (PEs) (6). The pattern of molecular species was similar in diacylglycerols and PCs (Fig. 1), indicating the absence of a significant substrate specificity for choline phosphotransferase against the major diacylglycerols. Hexaenoic diacylglycerols were preferentially utilized for PE synthesis, which has been observed by several workers (6–9).

In order to study this pathway in more detail, the metabolism of molecular species was studied *in vitro.* Experiments were performed with isolated hepatocytes, which were prepared by perfusion of rat liver with collagenase and hyaluronidase according to Berry and Friend (10) with the modifications described elsewhere (11). The cells were suspended in Hanks' solution containing 1.25-mM Ca^{2+} and 10-mM phosphate, pH 7.4. The most marked effect of albumin-bound oleic acid on glycerolipid synthesis from [³H]glycerol was a stimulation of triacylglycerol synthesis (Fig. 2) in accordance with studies in liver slices (12). Also the synthesis of different molecular species was drastically changed upon

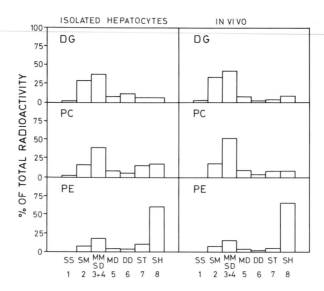

FIG. 1. Comparison of molecular species of glycerolipids synthetized from [³H]glycerol in rat liver *in vivo* (6) and in isolated rat hepatocytes (13). **S,** Saturated; **M,** monoenoic; **D,** dienoic; **T,** tetraenoic; **H,** hexaenoic; **DG,** diacylglycerols; **PC,** phosphatidylcholines; **PE,** phosphatidylethanolamines.

addition of increasing concentration of oleic acid (13). Among diacyl lipids the proportion of the monoenoic-monoenoic species increased and in triacylglycerols the trienoic species. In diacyl lipids the two species containing one oleic acid residue per molecule were elevated at the lower oleic acid concentrations but then decreased. The pattern of molecular species synthetized by hepatocytes in the absence of added fatty acid was very similar to that observed *in vivo* (Fig. 1).

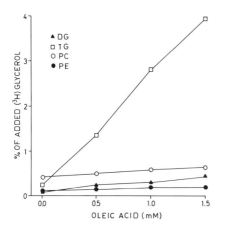

FIG. 2. Effect of oleic acid on glycerolipid synthesis from [³H]glycerol. Isolated hepatocytes (2.1 mg protein), 2-mM [³H]glycerol and oleic acid bound to 2% bovine serum albumin were incubated in Hanks' solution for 60 min. TG, triacylglycerols (□), DG (▲), (○) PC, (●) PE.

The effects of other fatty acids on the synthesis of molecular species was studied in similar experiments with labeled glycerol and isolated hepatocytes (13). Since diacylglycerols are the precursors of triacylglycerols, PCs, and PEs, a nonspecific utilization of diacylglycerols would be expected to give the same distribution of molecular species in all diacyl lipids. Addition of linoleic acid resulted in large amounts of the dilinoleoyl species in all three lipids, indicating a relatively non-specific utilization of this diacylglycerol for phospholipid synthesis. After the addition of stearic acid, a large proportion of diacylglycerols was fully saturated (Fig. 3). The proportion of saturated species in the phospholipids was considerably lower, indicating a low utilization of this species for phospholipid synthesis.

Experiments performed with other saturated fatty acids showed that saturated diacylglycerols were formed in considerable amounts in all cases (Table 1). They were utilized for PE synthesis to a low extent, whereas the utilization for PC synthesis depended upon chain length. Dimyristoyl diacylglycerol was well utilized, but the utilization decreased with decreasing and increasing chain length. Most probably the low utilization of saturated diacylglycerols is a reflection of the substrate specificities of choline phosphotransferase (EC 2.7.8.2.) and ethanolamine phosphotransferase (EC 2.7.8.1.). This may represent one of the mechanisms whereby the formation of phospholipid molecules with unsuitably high transition temperature is avoided. Some earlier studies on this problem in cell-free systems are hard to interpret, since the specificities observed were not as marked as those observed *in vivo* (7). When microsome-bound diacylglycerols were used, the specificities were higher (Kanoh, *this volume*). The advantage of

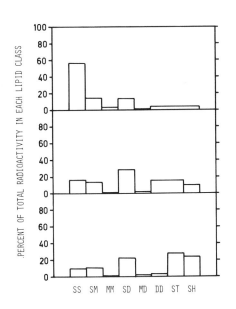

FIG. 3. Effect of stearic acid (1 mM) on the proportion of molecular species synthetized from [³H]glycerol, (upper) 1,2-diacyl-*sn*-glycerols, (center) PCs, (lower) PEs.

TABLE 1. *Effect of different saturated fatty acids on the synthesis of saturated glycerolipids from [³H]glycerol*

| Fatty acid added (1 mM) | Percent totally saturated species in each lipid class | | |
	Diacyl-glycerols	PCs	PEs
None	10.1	1.1	0.3
Lauric acid	64.6	25.6	3.7
Myristic acid	74.5	58.0	4.7
Palmitic acid	72.6	40.7	5.0
Stearic acid	55.4	15.5	9.7

Isolated rat hepatocytes (2×10^6) were incubated with 50 nmoles of [³H]glycerol and 2% albumin with bound fatty acid in Hanks' buffer containing 10-mM phosphate. Diacylglycerols, PCs, and PEs were isolated by thin-layer chromatography and then fractionated into different molecular species by thin-layer chromatography on $AgNO_3$-containing silica gel. The data indicate the proportion of totally saturated species within each lipid class.

the present experimental approach is that the utilization of diacylglycerols formed in the intact cell is studied.

There is evidence that diacylglycerols arise from sources other than phosphatidic acids, primarily from PCs through the reversal of the choline phosphotransferase reaction (14, 15). Until now most estimates on the quantitative importance of this reaction have been based on the specific radioactivities of choline-containing compounds in liver after the injection of labeled choline (14, 16–18; also Sundler and Åkesson, ref. 26). We have tried another approach to evaluate the role in phospholipid synthesis of diacylglycerols not originating from phosphatidic acids (19).

Phosphate-32 phosphate is incorporated into PCs and PEs via cytidine diphosphate (CDP) derivatives and reflects the reaction of all diacylglycerols in these pathways. Hydrogen-3-glycerol incorporated into PCs and PEs on the other hand reflects only the diacylglycerols, which are newly synthetized from phosphatidic acids. If all diacylglycerols originate from phosphatidic acids, the distribution of [³H]glycerol and [³²P] among molecular species of PCs and PEs would be expected to be the same. In contrast we observed significant differences when [³H]glycerol, [³²P]phosphate and albumin-bound oleic acid or linoleic acid were incubated with hepatocytes. The monoenoic-monoenoic (MM) or dienoic-dienoic (DD) species attained the highest radioactivity of both isotopes in PCs as well as PEs. The percentage of [³H]glycerol was, however, consistently larger than that for [³²P]phosphate, indicating that all diacylglycerols taking part in phospholipid synthesis do not originate from phosphatidic acids.

TABLE 2. *Calculated incorporation of different molecular species of diacylglycerols not originating from phosphatidic acids into PEs and PCs*

Molecular species	PEs				PCs		
	No fatty acid	18:1	18:2	Mean ($n = 6$)	18:1	18:2	Mean ($n = 9$)
1 (SM)	—	0.0	1.10	0.33	8.52	11.00	10.17
2 (MM)	0.0	—	0.11	0.04	—	2.35	1.57
3 (SD)	0.95	2.78	3.26	2.33	22.77	32.20	29.06
4 (MD)	0.0	0.0	0.94	0.31	6.96	10.56	9.36
5 (DD)	4.68	4.64	—	3.11	5.08	—	1.69
6 (ST)	30.78	24.44	19.28	24.84	35.55	27.36	30.09
7 (>4)	63.59	68.14	75.31	69.01	21.12	16.53	18.06

The calculations, which are described elsewhere (19), were performed from data obtained after incubation of hepatocytes with [^{32}P]phosphate, [^{3}H]glycerol, and either oleic acid, linoleic acid, or no fatty acid.

It was assumed that the major species formed in the presence of oleic acid or linoleic acid could be attributed exclusively to synthesis via phosphatidic acids. It could then be calculated that 26 and 13% of the diacylglycerols incorporated into PEs and PCs, respectively, originated from sources other than phosphatidic acids. The calculated fraction was constant on the addition of different fatty acids. The calculated composition of such diacylglycerols not originating from phosphatidic acids, which were incorporated into PEs and PCs, was also remarkably constant (Table 2). These diacylglycerols should therefore originate from a glycerolipid pool, which has a constant fatty acid composition during short-term incubation with oleic acid or linoleic acid. The composition of the diacylglycerols not originating from phosphatidic acids could not be directly determined. Instead we assumed that their incorporation into PCs occurred randomly due to the lack of specificity of choline phosphotransferase. To find the origin of these diacylglycerols, their distribution among PCs was compared with the mass composition of 1,2-diacyl-*sn*-glycerol units in the major glycerolipids of rat liver (Fig. 4). A striking similarity was found only with PCs, which strongly suggests that they originate from this lipid class. In summary these experiments indicate that diacylglycerols formed from other sources than phosphatidic acids take part both in PE and PC synthesis and that such diacylglycerols are formed by the reversal of the choline phosphotransferase reaction.

PHOSPHOLIPID BIOSYNTHESIS VIA BASE EXCHANGE

Another reaction, the quantitative importance of which has been debated, is the base exchange (20). For example [^{3}H]ethanolamine will exchange with another phospholipid base to yield labeled PEs. In these studies a method for the separation of PEs into molecular species that did not involve splitting off the ethanolamine portion was needed. Therefore PEs were treated with benzoylchlo-

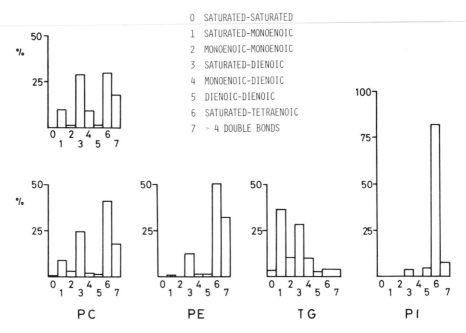

FIG. 4. Comparison of the calculated composition of diacylglycerols not originating from phosphatidic acids (top) with literature data on the mass composition of rat liver PCs (8), PEs (8), triacylglycerols (25), and phosphatidylinositols (PIs) (8). 0, saturated-saturated; 1, saturated-monoenoic; 2, monoenoic-monoenoic; 3, saturated-dienoic; 4, monoenoic-dienoic; 5, dienoic-dienoic; 6, saturated-tetraenoic; 7, > 4 double bonds.

ride and diazomethane, which gave N-benzoyl, O-methyl-PE (21). This derivative was separated by argentation chromatography, which yielded the same high resolution as the separation of different diacylacetylglycerols.

The relative rate of incorporation of labeled ethanolamine *via* base exchange and the CDP-ethanolamine pathway in isolated hepatocytes was evaluated in two different ways (22). First, it was assumed that ethanolamine incorporation remaining after total inhibition of energy-dependent PE synthesis by cyanide and fluoride was due to base exchange (Table 3). When isolated hepatocytes were preincubated with cyanide and fluoride prior to the addition of isotopes, PE synthesis from [^{32}P]phosphate was inhibited by more than 99%, whereas part of the incorporation of [^{3}H]ethanolamine remained. Second, we made use of the previous finding that addition of albumin-bound oleic or linoleic acid changes PE synthesis via CDP-ethanolamine so that high proportions of dioleoyl- or dilinoleoyl-PE are formed. Then the labeling of different PEs from [^{32}P]phosphate, which is incorporated only *via* CDP-ethanolamine, was compared with the labeling from [^{3}H]ethanolamine, which in addition is incorporated by base exchange. From this comparison the relative role of the two pathways could be calculated. If [^{3}H]ethanolamine and [^{32}P]phosphate are both incorporated only

TABLE 3. *Energy-independent incorporation of [³H]ethanolamine into PEs in isolated hepatocytes*

Incubation time (min after ^{32}P addition)	Preincubation with NaCN and NaF	Incorporation into PEs	
		[^{32}P]phosphate cpm $\times 10^{-3}$	[^{3}H]ethanolamine cpm $\times 10^{-6}$
30	−	157 (100)	1.034 (100)
45	−	317 (100)	1.760 (100)
30	+	0.540 (0.3)	0.093 (9.0)
45	+	0.535 (0.2)	0.189 (10.7)

The hepatocytes were preincubated for 30 min with or without 20-mM NaCN and 20-mM NaF prior to the addition of [^{32}P]phosphate. [^{3}H]ethanolamine was added 15 min after [^{32}P]phosphate addition. Data within parentheses indicate the incorporation expressed as percent of that in the absence of NaCN and NaF. For further details see ref. 22.

via CDP-ethanolamine, one would expect their distribution among the molecular species of PE to be identical. Instead it was found that the addition of linoleic acid (Fig. 5) or oleic acid affected the distribution of the two isotopes in a similar, but not identical, way. The difference between the distribution of the isotopes was essentially the same whether oleic or linoleic acid was added, and it was also constant at different times of incubation. The results therefore indicate that part of the labeled ethanolamine found in PEs had been incorporated by a mechanism other than the CDP-ethanolamine pathway responsible for [^{32}P]phosphate incorporation.

The fraction of ethanolamine incorporated via base exchange was calculated

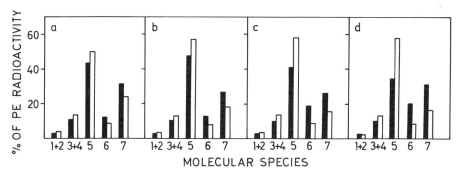

FIG. 5. Distribution of [^{32}P]phosphate (open bars) and [^{3}H]ethanolamine (black bars) among molecular species of PEs as a function of the initial concentration of ethanolamine (**a,** 0.03 mM; **b,** 0.06 mM; **c,** 0.17 mM; **d,** 0.41 mM). Isolated hepatocytes were incubated with [^{3}H]ethanolamine, [^{32}P]phosphate, 1.5-mM linoleic acid and 2% albumin in Hanks' solution for 45 min. Molecular species: 1 + 2, saturated plus monoenoic; 3 + 4, dienoic plus trienoic; 5, dienoic-dienoic; 6, saturated-tetraenoic; 7, species with more than four double bonds. (Reproduced from ref. 22 with permission.)

TABLE 4. *Increase in the quantitative role of base exchange in isolated hepatocytes as a function of ethanolamine concentration*

Ethanolamine concentration (mM)	A	B
0.01	9.4	7.7
0.03	12.2	12.8
0.06	21.4	16.4
0.17	24.1	29.4
0.41	30.3	40.0

The percentage contribution of base exchange was calculated from the energy-independent incorporation of [^{3}H]ethanolamine (A) or from data obtained after incubation with linoleic acid (B) (see Fig. 5). For further details see ref. 22.

by the two methods (Table 4). Results obtained using the two approaches agreed very well and showed that at a physiologic concentration of ethanolamine (0.02 mM, *unpublished results*) 8 to 9% of its total incorporation into PEs can be attributed to base exchange. That this additional incorporation occurred via the calcium-stimulated base exchange is further supported by the fact that ethylene diamine tetra-acetic acid (EDTA) and an excess of serine were inhibitory (22). Furthermore the incorporation of ethanolamine into different PE species by base exchange in hepatocytes was in reasonable agreement with characteristics of the calcium-stimulated incorporation of ethanolamine in rat liver microsomes found by Bjerve (23). When the concentration of [^{3}H]ethanolamine in the incubation medium was raised, the difference in distribution of [^{3}H]ethanolamine and [^{32}P]phosphate among molecular species of PE was augmented (Fig. 5). This indicates that an increasing proportion of the labeled ethanolamine had been incorporated via base exchange (Table 4). Also the energy-independent proportion of the total [^{3}H]ethanolamine incorporation increased at higher concentrations of ethanolamine.

The relevance of results obtained by the present approach to the situation *in vivo* was tested by determination of the relative rate of choline exchange. A previous study indicated that the rate of this reaction is about 5% of that of the CDP-choline pathway in intact rat liver (18). In isolated hepatocytes the energy-independent incorporation of 0.04-mM [Me-^{3}H]choline was 4.5 and 4.9% of that in uninhibited control incubations in two separate experiments. We have concluded that base exchange occurs in isolated hepatocytes and that it has a low quantitative importance in PE and PC synthesis.

SECRETION OF LIPOPROTEINS BY ISOLATED HEPATOCYTES

In some experiments we investigated whether isolated hepatocytes secrete lipoproteins. After incubation the cells and the medium were separated by centrifugation. The percentage of total glycerolipid radioactivity recovered in the medium increased with time after incubation with [³H]glycerol (24). Most of the medium radioactivity was in triacylglycerols and most of the lipoproteins behaved as very low-density lipoproteins (VLDL) on ultracentrifugation and in several other tests. Then different fatty acids and [³H]glycerol were added to the incubation medium and the distribution of label among molecular species of triacylglycerol in VLDL and in the cells was analyzed (Fig. 6). When no fatty acid was added and when unsaturated fatty acids were added, the distribution was very similar in VLDL and in cells. On the other hand, the fully saturated species, which was abundant in the cells after the addition of saturated fatty acids, was secreted to a lower degree than unsaturated triacylglycerols. The mechanism behind such selective secretion is unknown but might be due to a less favorable packing of saturated triacylglycerols within the lipoprotein particle.

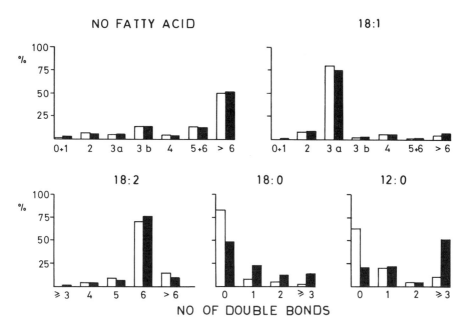

FIG. 6. Comparison of molecular species of triacylglycerols synthetized in hepatocytes and secreted by hepatocytes in VLDL. The incubations were performed as described earlier (24) and VLDL was isolated by ultracentrifugation. The separation of molecular species is described elsewhere (25). The data are expressed as percent of total triacylglycerol radioactivity. Black bars, VLDL-triacylglycerols; open bars, cell triacylglycerols.

CONCLUDING REMARKS

These studies have shown that enzymatically isolated hepatocytes are a useful experimental system for studies on glycerolipid metabolism. The biosynthesis of molecular species of lipids has many similarities to that *in vivo* and the relative role of different pathways can be studied. The relative synthetic rates for different molecular species and lipid classes can be manipulated to some degree. Through long-term incubations, the mass composition of cell phospholipids might be changed so that the effect of such changes on different membrane functions can be studied.

ACKNOWLEDGMENTS

This work was supported by grants from the Swedish Medical Research Council (project 03X-3968), H. Jeanssons Foundation, and A. Påhlssons Foundation.

REFERENCES

1. Demel, R. A., Kinsky, S. C., Kinsky, C. B., and van Deenen, L. L. M. (1968): *Biochim. Biophys. Acta,* 150:655.
2. De Gier, J., Mandersloot, J. G., and van Deenen, L. L. M. (1968): *Biochim. Biophys. Acta,* 150:666.
3. McElhaney, R. N., De Gier, J., and van der Neut-Kok, E. C. M. (1973): *Biochim. Biophys. Acta,* 298:500.
4. Machtiger, N. A., and Fox, C. F. (1973): *Ann. Rev. Biochem.,* 42:575.
5. Kennedy, E. P. (1957): *Ann. Rev. Biochem.,* 26:119.
6. Åkesson, B., Elovson, J., and Arvidson, G. (1970): *Biochim. Biophys. Acta,* 210:15.
7. Kanoh, H. (1970): *Biochim. Biophys. Acta,* 218:249.
8. Holub, B. J., and Kuksis, A. (1971): *Can. J. Biochem.,* 49:1347.
9. Vereyken, J. M., Montfoort, A., and van Golde, L. M. G. (1972): *Biochim. Biophys. Acta,* 260:70.
10. Berry, M. N., and Friend, D. S. (1969): *J. Cell Biol.,* 43:506.
11. Nilsson, Å., Sundler, R., and Åkesson, B. (1973): *Eur. J. Biochem.,* 39:613.
12. Rose, H., Vaughan, M., and Steinberg, D. (1963): *Am. J. Physiol.,* 206:345.
13. Sundler, R., Åkesson, B., and Nilsson, Å. (1974): *J. Biol. Chem.,* 249:5102.
14. Björnstad, P., and Bremer, J. (1966): *J. Lipid Res.,* 7:38.
15. Kanoh, H., and Ohno, K. (1973): *Biochim. Biophys. Acta,* 306:203.
16. Bjerve, K. S., and Bremer, J. (1969): *Biochim. Biophys. Acta,* 176:570.
17. Treble, D. H., Frumkin, S., Balint, J. A., and Beeler, D. A. (1970): *Biochim. Biophys. Acta,* 202:163.
18. Sundler, R., Arvidson, G., and Åkesson, B. (1972): *Biochim. Biophys. Acta,* 280:559.
19. Sundler, R., Åkesson, B., and Nilsson, Å. (1974): *Biochim. Biophys. Acta,* 337:248.
20. Borkenhagen, L. F., Kennedy, E. P., and Fielding, L. (1961): *J. Biol. Chem.,* 236:PC 28.
21. Sundler, R., and Åkesson, B. (1973): *J. Chromatogr.,* 80:233.
22. Sundler, R., Åkesson, B., and Nilsson, Å. (1974): *FEBS Lett.,* 43:303.
23. Bjerve, K. S. (1973): *Biochim. Biophys. Acta,* 296:549.
24. Sundler, R., Åkesson, B., and Nilsson, Å. (1973): *Biochem. Biophys. Res. Commun.,* 55:961.
25. Akesson, B. (1969): *Eur. J. Biochem.,* 9:463.
26. Sundler, R., and Åkesson, B. (1975): *Biochem. J.,* 146:309.

Lipids, Vol. 1, edited by R. Paoletti, G. Porcellati, and G. Jacini. Raven Press, New York © 1976.

Formation of Molecular Species of Phosphatidylcholine and Phosphatidylethanolamine by Acylation of Endogenous Lysophospholipids in Rat Liver Extracts

L. Gosselin

University of Liège, Laboratory of Medical Chemistry, B-4000 Liège, Belgium

Fatty acid introduction in a given phospholipid species may proceed either *via* the Kennedy pathway or *via* the acylation of a lysophosphatide. In this study, by adding [^{14}C]-labeled fatty acids together with tritiated glycerol to concentrated homogenates of rat liver and comparing the extent of the incorporation of both radioactive tracers in several molecular species of lecithin, it was observed that the introduction of palmitic acid into the tetraenoic and disaturated lecithins might partly occur *via* the acylation of a preexisting lysolecithin that apparently was not generated in the medium. On the other hand, experimental evidence was obtained that suggested that the uptake of palmitic acid into some molecular species of phosphatidylethanolamine (PE) might depend, at least to some extent, on a generation of a 2-monoacyl derivative of PE.

INCORPORATION OF SATURATED FATTY ACIDS IN VARIOUS MOLECULAR SPECIES OF LECITHIN IN RAT LIVER HOMOGENATES

The first series of experiments were designed to study the biosynthesis of the various molecular species of lecithin in rat liver homogenates or in rat liver slices (1, 2). The preparation of concentrated homogenates of rat liver and the composition of the incubation medium were as described by Foster and Bloom (3); such a system is capable of carrying out the synthesis of fatty acids from acetate as well as the *de novo* synthesis of phospholipids. The fractionation of the molecular species of lecithin was performed by phospholipase C (E.C. 3.1.4.3), degradation of the purified lecithins (4), and resolution of the resulting diglycerides mixture by thin-layer chromatography (TLC) on plates coated with silica gel impregnated with AgNO$_3$ (25%, w/w) according to van Golde and van Deenen (5). The phospholipase C from *Clostridium welchii* was supplied by Sigma Chemical Co. (US).

In the experiment described in Fig. 1, the simultaneous incorporation of ^{14}C-labeled fatty acids and 1(3)-[^{3}H]glycerol in lecithin was studied at two time intervals. The following was observed:

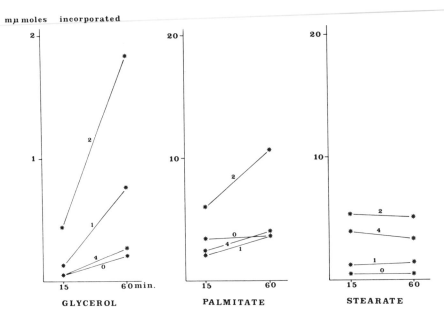

FIG. 1. Simultaneous incorporation of [¹⁴C]-labeled fatty acids and 1(3)-[³H]glycerol into 4 molecular species of lecithin by a liver homogenate. (Reproduced from Gosselin, ref. 2.)

(a) The amount of fatty acids introduced *via* the Kennedy pathway increased considerably between the 15th and the 60th min of the incubation. On the other hand, the amount of fatty acids introduced *via* the acylation of the endogenous lysolecithins did not increase after the 15th min owing possibly to a limitation of the endogenous acceptor.

(b) The introduction of stearic acid in any species of lecithin seemed to occur solely *via* the acylation of lysolecithins. The incorporation of saturated fatty acids via the acylation of glycerol-3-phosphate resulted mainly in the formation of a palmitoyl-linoleyl species and, to a lesser extent, of palmitoylarachidonyl and palmitoyloleyl species. These findings are consistent with other authors' conclusions (6–9) and in particular with conclusions drawn from experiments performed *in vivo* (7–9).

(c) The relative importance of palmitate incorporation into a disaturated lecithin could be ascribed to the relative abundance of the 1-acyl-*sn*-glycero-3-phosphorylcholine (10), assuming that the positional specificity was not absolute (11).

The experiment recorded in Table 1 was designed to compare the pathway of introduction of the fatty acids added to the medium with that of the fatty acids neosynthesized during the course of incubation. This was done by incubating simultaneously 1(3)-[³H]glycerol with either a saturated fatty acid labeled with ¹⁴C or with 1,2-[¹⁴C]acetate and by determining the ³H/¹⁴C ratio in several molecular species of lecithin at the end of a 60-min incubation. The radioactive

TABLE 1. *Incorporation of neosynthesized or added [¹⁴C]-fatty acids into various lecithins compared to that of 1(3)-[³H] glycerol in a liver homogenate*

Molecular species (number of double bonds)	$^3H/^{14}C$ ratio		
	1,2-[¹⁴C] acetate	1-[¹⁴C] palmitate	1-[¹⁴C] stearate
5–6	74.89	3.00	3.32
4	65.74	1.65	1.58
0 + 3	71.48	4.04	3.47
1 + 2	74.16	15.24	31.14
2	62.91	4.70	6.07
1	72.72	3.73	8.42
0	52.75	1.18	5.28

fatty acids derived from acetate and introduced subsequently into lecithin consisted mainly of palmitic acid.

When the [¹⁴C]palmitate or [¹⁴C]stearate were added in the medium, the $^3H/^{14}C$ ratio was found to be lower in the tetraenoic species than in the other lecithins reflecting thereby the relative importance of the lysophosphatide pathway for the synthesis of this particular species (with palmitic acid, a low ratio was also noted in the disaturated species). The lowering of the $^3H/^{14}C$ ratio in the tetraenoic species could not be observed, however, when labeled acetate was used as the fatty acid precursor; this suggested that the palmitic acid synthesized *in situ* could only be introduced in lecithin *via* the Kennedy pathway.

By incubating simultaneously tritiated glycerol and [¹⁴C]-labeled acetate with liver slices instead of homogenates, it was also repeatedly observed that the value of the $^3H/^{14}C$ ratio found for the different molecular species of lecithin was remarkably constant. A possible explanation of these findings is that the neosynthesized palmitate can only enter the lecithin molecule via the Kennedy pathway because there is no more lysolecithin available for acylation, assuming that the acylation of lysolecithins by the endogenous fatty acids is more rapid than the synthesis of fatty acids and also that the pool of pre-existing lysolecithins cannot be replenished *in vitro*.

INCORPORATION OF SATURATED FATTY ACIDS INTO PE IN A RAT LIVER HOMOGENATE

Tritiated glycerol and [¹⁴C]-labeled palmitate or acetate were simultaneously incubated with a liver homogenate exactly as described in the experiment recorded in Table 1 and at the end of the incubation the PE were extracted, purified by TLC, and resolved into several molecular species. As the commercially available phospholipase C from *Cl. welchii* did not hydrolyze PE, the direct subfractionation of the intact molecules was achieved by TLC on silica gel impregnated with AgNO₃ using Arvidson's method (12), modified as follows: precoated plates

TABLE 2. *Simultaneous incorporation into PE of 1(3)-[^{3}H]glycerol and 1-[^{14}C]palmitate or 1,2[^{14}C]acetate in a liver homogenate*

Molecular species (number of double bonds)	^{3}H/^{14}C ratio	
	Palmitate	Acetate
6	0.78	9.00
5	0.82	8.70
4 + 1	1.68	15.70
4	0.49	5.25
2	1.03	13.30

of silica gel (Merck, 250-μ thickness) were sprayed with a 15% aqueous solution of AgNo$_3$ (about 10 ml/200 cm^2) and dried for 1 hr at room temperature in the dark; the plates were then left for 2 hr in an oven at 100°C and used immediately after activation. Resolution of five or six molecular species was achieved by using the system chloroform/methanol/acetic acid/water (50/20/4/2,v/v) as the developing solvent; the detection and elution were performed as in the original method. This allowed the isolation of the three main species of PE (hexaenoic, tetraenoic, and dienoic), as well as the separation of two minor components tentatively identified as pentaenoic species on the basis of fatty acid analysis. The dienoic species was usually contaminated by traces of a monoenoic species.

The results of the experiment (Table 2) show that the ^{3}H/^{14}C ratio was lower in the tetraenoic species than in the hexaenoic and dienoic when the labeled fatty acids either were added to the medium or were synthesized *in situ* from the labeled acetate. This suggested that some of the neosynthesized palmitate could be introduced in PE via the acylation of some endogenous lysophosphatides.

FATTY ACIDS INCORPORATION INTO PHOSPHOLIPIDS IN ISOLATED MICROSOMES

The incorporation of fatty acids into phospholipids in a system of isolated particles that is not capable of carrying out the *de novo* synthesis of phospholipids (13–15) must occur via the acylation of endogenous lysoderivatives and will be limited by the amount of these acceptors (11, 16). Trace amounts of 1-[^{14}C]palmitic or 1-[^{14}C]linoleic acid have been incubated with liver microsomes in a standard medium containing 15 mg of microsomal proteins, 300 μmoles of KCl, 40 μmoles of Tris adjusted with HCl to pH 7.4, 5 μmoles of adenosine triphosphate (ATP), 5 μmoles of MgCl$_2$, and 100 nmoles of coenzyme A (Co A) in a final volume of 2 ml. All incubations were carried out at 37°C. The microsomes

TABLE 3. *Palmitate incorporation in lecithin (PC) and PE at acid pH values (in percent of incorporation at pH 7.4)*

	Experiment								
	1			2			3	4	5
	pH 6.8	pH 6.3	pH 5.8	pH 6.8	pH 6.0	pH 5.4	pH 5.6	pH 5.4	pH 5.4
PC	101	105	99	106	142	73	171	66.3	59.4
PE	101	85	55	68	66	11	29	8.5	14.9

were obtained from a 10% homogenate in 0.25-M sucrose as the particles sedimenting between 15,000 X g (15 min) and 100,000 X g (90 min).

The fatty acid incorporation into lecithin and PE proceeded fairly rapidly, and after 15 min of incubation no further incorporation occurred. Over 95% of the radioactive linoleic acid was incorporated at position 2 of lecithin, indicating thereby the good positional specificity of the enzyme system. However, 20% of the palmitate incorporated into lecithin was found in a disaturated species; this resulted probably from the relative abundance of the endogenous 1-acyl-*sn*-glycero-3-phosphorylcholine. Linoleic acid was incorporated predominantly into lecithin, whereas palmitic acid transfer occurred to the same extent in lecithin and in PE.

As shown in Table 3, the pattern of the palmitate incorporation into phospholipids was influenced by pH. A lowering of the pH of the incubation medium resulted in a selective impairment of the palmitate incorporation into PE. Since the optimal pH of the microsomal phospholipase A1 is in the alkaline range (17, 18), this might be taken to indicate that the palmitic acid uptake in PE could depend to some extent on a continuous supply of lyso-PE. Other explanations, however, cannot be ruled out.

PRODUCTION OF LYSO-PE IN LIVER MICROSOMES *IN VITRO*

The occurrence in liver tissue of both structurally isomeric lysolecithins has been demonstrated (10); however, to our knowledge, no analogous study on isomeric lyso-PEs has ever been performed. The experimental findings reported above prompted an investigation of the lyso-PE content of the liver microsomes. It was of particular interest to determine the type and the amount of these lyso derivatives present in the nonincubated as well as in the incubated particles.

The microsomal lipids from at least 2 g of liver were transferred on a small column of DEAE cellulose (1.6 X 3.5 cm), and the column was eluted with four volumes of chloroform/methanol (9/1) followed by four volumes of methanol. The lysolecithins and the lyso-PEs can then be easily isolated by fractionating the first and the second eluate, respectively, on a thin layer of silica gel using the

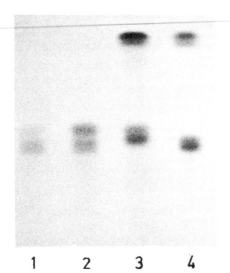

FIG. 2. Production of lyso-PE in incubated microsomes. 1,lyso-PE from nonincubated microsomes; 2,lyso-PE from microsomes incubated at pH 7.4; 3, 2-monoacyl-PE dylethanolamine (lower spot); 4, 1-monoacyl-PE (lower spot). Solvent system: chloroform/methanol/water (65/25/4). Detection by the ninhydrine reagent.

1   2   3   4

system chloroform/methanol/acetic acid/water (50/25/6.6/3.3,v/v) as the developing solvent. Incubation of the microsomes for 30 min at 37°C in the standard medium (pH 7.4) from which the ATP and the Co A had been omitted did not lead to any significant change of the lysolecithin concentration. Such treatment, however, greatly increased the lyso-PE content of the particles (Figs. 2 and 3). Separation of the 1-acyl and 2-acyl isomers of lyso-PE was achieved by using precoated plates of silica gel (Merck). The 2-acyl isomer used as the reference compound was prepared from beef heart PE (19); the 1-acyl isomer could be easily obtained by hydrolyzing liver PE with the *Crotalus adamanteus* venom (20). In the solvent system chloroform/methanol/water (65/25/4,v/v), Rf values for the 1-acyl and 2-acyl isomers were found to be 0.33 and 0.37, respectively.

When purified lipid fractions obtained from incubated and nonincubated microsomes were chromatographed in that system, it was shown that both types of particles contained substances reacting with ninhydrin, staining positively

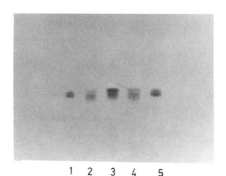

FIG. 3. Influence of the pH on the production of lyso-PE in microsomal particles incubated at 37°C. 1 and 5,lyso-PE prepared by treating PE with snake venom; 2, lyso-PE from nonincubated particles; 3, lyso-PE from particles incubated at pH 7.4; 4, lyso-PE from particles incubated at pH 5.4. Solvent system and visualization as in Fig. 2.

1   2   3   4   5

TABLE 4. *Fatty acid composition of two isomeric lyso-PEs (in wt%)*

Structure	Lower spot[a]	Upper spot[b]
16:0	36.8	18.6
18:0	13.8	0
16:1	0	6.0
18:1	33.0	3.0
18:2	16.4	31.7
20:4	0	23.2
22:6	0	17.5

[a] Recovered from nonincubated microsomes.
[b] Recovered from microsomes incubated at pH 7.4.

with the molybdenum blue reagent for phospholipids (21), and exhibiting chromatographic properties similar to those of the isomeric lyso-PEs. It could also be shown that the lipid accumulating in the incubated particles had a chromatographic behavior similar to that of the 2-monoacyl PE (Fig. 2) and was not formed in substantial amount when the incubation was carried at pH 5.4 instead of pH 7.4 (Fig. 3).

The slow- and fast-moving derivatives were isolated from microsomal extracts on a preparative scale and purified by TLC on precoated plates of silica gel. Both substances were shown to contain ethanolamine as the main water-soluble product after acid hydrolysis. Fatty acid analysis (Table 4) showed that the substance having the higher R_f value contained mainly polyunsaturated fatty acids, whereas the substance having the lower R_f value was found to contain mostly saturated and monoenoic acids. It thus appeared that the product accumulating during the incubation of the particles at pH 7.4 was indeed the 2-acyl-*sn*-glycero-3-phosphorylethanolamine.

SUMMARY

In liver tissue *in vitro,* the palmitic acid may be introduced in a tetraenoic lecithin partly via the acylation of a pre-existing lysolecithin, which does not seem to be generated in situ. The possibility that most of the liver lysolecithins available for acylation might have an extrahepatic origin should be kept in mind. On the other hand, the transfer of palmitate into PE might very well depend on a continuous supply of lyso-PE, at least at the beginning of the incubation. These suggestions are in keeping with the well-known fact that the phospholipases A of the liver particles preferentially hydrolyze PE (17, 22–24).

ACKNOWLEDGMENTS

The skilled technical assistance of Mr. R. Descamps and Mrs. J. Viana de Carvalho Buraca is gratefully acknowledged.

REFERENCES

1. Gosselin, L. (1972): *Arch. Int. Physiol. Biochim.,* 80:189.
2. Gosselin, L. (1972): *Rev. Ferment. Indust. Aliment.,* 27:71.
3. Foster, D. W. and Bloom, B. (1963): *Biochim. Biophys. Acta,* 70:341.
4. Renkonen, O. (1966): *Biochim. Biophys. Acta,* 125:288.
5. van Golde, L. M. G., and van Deenen, L. L. M. (1966): *Biochim. Biophys. Acta,* 125:496.
6. van Golde, L. M. G., Scherphof, G. L., and van Deenen, L. L. M. (1969): *Biochim. Biophys. Acta,* 176:635.
7. Åkesson, B., Elovson, J., and Arvidson, G. (1970): *Biochim. Biophys. Acta,* 210:15.
8. Åkesson, B., Elovson, J., and Arvidson, G. (1970): *Biochim. Biophys. Acta,* 218:44.
9. Åkesson, B. (1970): *Biochim. Biophys. Acta,* 218:57.
10. van den Bosch, H., and van Deenen, L. L. M. (1965): *Biochim. Biophys. Acta,* 106:326.
11. van den Bosch, H., van Golde, L. M. G., Eibl, H., and van Deenen, L. L. M. (1967): *Biochim. Biophys. Acta,* 144:613.
12. Arvidson, G. A. E. (1968): *Eur. J. Biochem.,* 4:478.
13. Lands, W. E. M. (1960): *J. Biol. Chem.,* 235:2233.
14. Lands, W. E. M., and Hart, P. (1965): *J. Biol. Chem.,* 240:1905.
15. Possmayer, E., Scherphof, G. L., Dubbelman, T. M. A. R., van Golde, L. M. G., and van Deenen, L. L. M. (1969): *Biochim. Biophys. Acta,* 176:95.
16. Sarzala, M. G., van Golde, L. M. G., DeKruyff, B., and van Deenen, L. L. M. (1970): *Biochim. Biophys. Acta,* 202:106.
17. Bjornstad, P. (1966): *Biochim. Biophys. Acta,* 116:500.
18. Newkirk, J. D., and Waite, M. (1971): *Biochim. Biophys. Acta,* 225:224.
19. Frosolono, M. F., Slivka, S., and Charms, B. L. (1971): *J. Lipid Res.,* 12:96.
20. Lands, W. E. M., and Merkl, I. (1963): *J. Biol. Chem.,* 238:898.
21. Dittmer, J. C., and Lester, R. L. (1964): *J. Lipid Res.,* 5:126.
22. Scherphof, G. L., and van Deenen, L. L. M. (1965): *Biochim. Biophys. Acta,* 98:204.
23. Bjornstad, P. (1966): *J. Lipid Res.,* 7:612.
24. Waite, M., and van Deenen, L. L. M. (1967): *Biochim. Biophys. Acta,* 137:498.

Lipids, Vol. 1, edited by R. Paoletti, G. Porcellati, and G. Jacini. Raven Press, New York © 1976.

Studies on the Specificity of the Reactions of Glycerol Lipid Biosynthesis in Rat Liver Using Membrane-Bound Substrates

H. J. Fallon, J. Barwick, R. G. Lamb, and H. van den Bosch

Department of Internal Medicine, Medical College of Virginia, Richmond, Virginia 23298 and Laboratory of Biochemistry, State University of Utrecht, Utrecht, Netherlands

Previous studies have described the enzymatic pathways of hepatic glycerolipid biosynthesis (1–4). Estimates of the specificity of these reactions for lipid substrates of various molecular species have been made by several techniques. The results in intact animals, liver slices, or in microsomal preparations have indicated a high degree of specificity in the formation of phosphatidate from *sn*-glycerol-3-phosphate (5–7). Thus, in mammalian systems saturated fatty acid species appear predominately in position 1 of phosphatidate, and the more unsaturated fatty acid are largely found in position 2. Although the substrate specificity of these reactions is only partial, it is sufficient to account for the unique distribution of fatty acid species found in liver phosphatidates (8). Similar studies have described relatively little specificity for phosphatidate phosphohydrolase or in the acylation of diglyceride (DG) to triglyceride (TG). Although the fatty acid distribution of rat liver TGs differs somewhat from that observed in DG synthesized *de novo,* this difference has been attributed to the formation of TG derived from DG formed by a reversal of the choline phosphotransferase reaction (9). Since it is well known that the fatty acid distribution in lecithin is different from that found in the neutral glycerides, this difference could account for these observations. However, direct measurement of the specificity of either diglyceride acyltransferase or choline phosphotransferase has been difficult because of the use of lipid substrates in nonphysiologic micellar or aqueous dispersions and the studies have given conflicting results.

Therefore we have developed methods to measure the individual reactions in hepatic glycerolipid biosynthesis utilizing labeled substrates formed endogenously in the microsomal membrane. Presumably such substrates should give information regarding these reactions which more closely resemble those obtained *in vivo.* This chapter describes the techniques for preparation of such substrates and the use of these preparations to study the specificity of the individual reactions in the pathway.

METHODS AND MATERIALS

Preparation of Microsomes

Male Wistar rats weighing 200 to 250 g and maintained on laboratory chow were utilized for the preparation of microsomes by previously described techniques (3, 5).

Preparation of Microsomal-Bound Substrates

Approximately 2.5 mg of microsomal protein were incubated in 1.7 ml of 35-mM Tris-HCl, pH 8.3, containing 6.25 mg of fatty-acid-poor albumin, 0.15-mM dithiothreitol, 0.3-mM *rac* glycerol-3-phosphate (GPA), 0.7 μCi *sn*-[1,3^{14}C]-GPA, 60-mM NaF, 50-mM palmitoyl-coenzyme A (CoA), and 50-mM oleoyl-CoA. This mixture was incubated for 20 min at 37°C and an additional 50-nmoles palmitoyl-CoA and 50-nmoles oleoyl-CoA added in a volume of 0.1 ml. The incubation was continued for 20 min longer and stopped by placing the mixture in ice. The incubation mixture was then placed over 5 ml of 0.3-M sucrose containing 0.1-mM dithiothreitol and resedimented at 80,000 $\times$ g in a number 40 rotor for 60 min in a Beckman ultracentrifuge. The resulting pellet was washed with 0.3-M sucrose and carefully homogenized in 0.7 ml of 0.3-M sucrose containing dithiothreitol. In a usual incubation mixture, 9.3-nmoles of phosphatidate containing 60,500 dpm were formed per milligram of microsomal protein. This microsomal-bound [^{14}C]phosphatidate was then further incubated with a rat liver cytosol preparation containing phosphatidate phosphohydrolase activity as described previously (3). Approximately 50% of the remaining radioactive phosphatidate was converted to DG by this technique.

Diglyceride Acyltransferase

The microsomal bound [^{14}C]DG (0.1- to 0.3-mg protein) was incubated in a 0.45-ml volume of 0.05-M K-phosphate buffer, pH 6.5, containing 2-mM dithiothreitol, 2.5-mg fatty-acid-poor albumin, 0.1-mM Mg Cl$_2$, 50-mM NaF, and 70 nmoles of various acyl CoA derivatives. The mixture was incubated for 1 to 4 minutes at 37°C and the reaction stopped by addition of 1.5 ml of 2:1 methanol-chloroform. Lipids were extracted as described below.

Choline Phosphotransferase

The microsomal-bound [^{14}C]DG (0.3- to 0.5-mg protein) was incubated in 0.2 ml of 0.1-M Tris-HCl buffer, pH 7.2, containing 10-mM MgCl$_2$, 50-mM β-mercaptoethanol, and 0.5-mM cytidine diphosphate choline (CDP-choline). The mixture was incubated at 37°C for 2 to 4 min and the lipids extracted as described below.

Lipid Extraction and Identifications

Lipids were extracted using a modification of the Bligh and Dyer Technique as described earlier (10). The reaction products were identified by thin-layer chromatography (TLC) using silica gel HR and a solvent system containing petroleum ether (BP 40–60), diethyl ether, and formic acid (60:40:1.5). The phospholipids were identified by TLC using solvent systems previously described (5, 10). For identification of molecular species, the DG regions were scraped from TLC plates and the lipid extracted from the silica gel with choloroform methanol (1:2 v/v). The species were separated by argentation chromatography with chloroform ethanol (97:3 v/v) as the solvent system. The DG standards were prepared by the degradation of egg lecithin, rat liver lecithin, and synthetic dipalmitoyl, dioleoyl, and dilinoleoyl-lecithin with phospholipase C. The molecular species of phosphatidate were identified following their elution from silica gel by formation of the dimethyl phosphatidates as described by Possmayer et al. (6). The molecular species of lecithin formed by the choline phosphotransferase reaction were identified following degradation with phospholipase C and separation as DG by the methods described above. Internal standards were added to assess losses during this procedure.

RESULTS

The previous observation indicating the specificity of the reactions leading to formation of phosphatidate from *sn*-GPA was confirmed in these studies. Ap-

TABLE 1. *Formation of various glycerolipid species from* sn-*[1,3^{14}C] GPA and palmitoyl-CoA by rat liver microsomes*

Molecular species[a]	Palmitoyl-CoA		
	PA[b]	DG[b]	PC[b]
0	64.5	84.5	53.5
1	5.3	4.4	5.2
1–1	5.0	0.4	4.1
2	17.0	10.0	21.5
1–2	3.3	0	4.1
poly	4.5	0.4	7.5

[a] The number of unsaturated bonds in lipid fatty acids.

[b] PA, phosphatidate; PC, phosphatidylcholine.

Values given indicate the percent distribution of radioactivity in each of the molecular species of lipid identified.

TABLE 2. *Formation of various glycerolipid species from sn-[1,3¹⁴C] GPA and a mixture of palmitoyl-CoA and oleoyl-CoA by rat liver microsomes*

Molecular species[a]	Palmitoyl-CoA + Oleoyl-CoA		
	PA	DG	PC
0	22.0	12.7	9.8
1	31.7	43.5	35.8
1–1	11.3	15.0	11.8
2	13.5	19.1	24.9
1–2	4.4	8.4	11.0
poly	2.9	1.2	6.5

[a] Refers to the number of unsaturated bonds in lipid fatty acids.

proximately 80% of saturated fatty acids were incorporated into position 1 of phosphatidate and 60 to 65% of unsaturated long chain fatty acids into position 2. These results conform to the molecular distribution of fatty acids in phosphatidates isolated from rat liver as described by several investigators (6, 11).

The results of studies of the distribution of palmitoyl-CoA in various molecular species of phosphatidate, DG, and phosphatidylcholine (PC) are shown in Table 1. It is apparent that phosphatidate species other than dipalmitoyl phosphatidate

TABLE 3. *Relative rate of phosphatidate formation and of phosphatidate phosphohydrolase activity with various acyl-CoA substrates*

Acyl-CoA	Relative rate	
	PA formation[a]	DG formation[b]
Palmitoyl	1.00	1.00
Palmitoleoyl	0.51	0.14
Stearoyl	0.72	0.74
Oleoyl	0.77	0.52
Palmitoyl + stearoyl	1.03	0.61
Palmitoyl + oleoyl	1.01	0.48
Palmitoyl + palmitoleoyl	1.07	0.22
Palmitoleoyl + stearoyl	0.35	0.23
Palmitoleoyl + oleoyl	0.73	0.04

[a] Refers to the relative rate of phosphatidate formation from sn-[1,3¹⁴C]GPA and the indicated acyl-CoA substrates with palmitoyl-CoA taken as 1.00. Incubation conditions did not favor DG formation.

[b] Incubation of microsomes containing phosphatidate formed in preceding reaction with Mg^{2+} and a supernatant fraction. The relative reaction is lower for all phosphatidates containing palmitoleate.

TABLE 4. *Various DG species formed using different acyl-CoA substrates*

| Molecular species[a] | DG Species | | | |
	Palmitoyl	Oleoyl	Palmitoyl oleoyl	Palmitoyl linoleoyl
0	84.5	17.3	12.7	11.4
1	4.4	26.5	43.5	7.9
1–1	0.4	32.2	15.0	6.1
2	10.0	12.8	19.1	62.0
1–2	0.0	10.0	8.4	2.9
poly	0.4	1.0	1.2	9.6

[a] The number of unsaturated bonds in lipid fatty acids.

are formed in this reaction. Since the palmitoyl-CoA substrate was chromatographically pure and there was no addition of ATP or CoA to the reaction mixture, it is likely that the microsomal preparations contained small amounts of other acyl-CoA derivatives, resulting in the formation of several molecular species of phosphatidate.

There was no remarkable difference in the conversion of these different species of phosphatidate to DG by the soluble phosphatidate phosphohydrolase preparation used. However, it is clear that the PC species formed from these DGs contained relatively less saturated species and relatively more diunsaturated fatty acids. Therefore, under these conditions, the choline phosphotransferase reaction evidences specificity toward diunsaturated species with less favorable rates for dipalmitoyl diglyceride. This is in contrast to previous findings in which aqueous dispersions of diglyceride species were used (12, 13). When a mixture of palmitoyl-CoA and oleoyl-CoA was used as substrate, a somewhat similar result was found as shown in Table 2. However, a much wider mixture of phosphatidate species were formed. There was no remarkable difference in the fatty acid content of the phosphatidate and DG formed under the reaction conditions. However, the PC formed contained less disaturated and more unsaturated species. This result is in keeping with the observed molecular distribution of fatty acids in rat liver lecithin. Similar results indicating specificity in this reaction were obtained when mixtures of palmitoyl-, oleoyl-, and linoleoyl-CoA were incubated with [14C]-labeled sn-GPA. This series of experiments would indicate that phosphatidate phosphohydrolase has little or no specificity toward various phosphatidate species, whereas considerable specificity in the formation of lecithins from DG was observed.

The specificity of phosphatidate phosphohydrolase was studied by an alternative method, and the results are shown in Table 3. Although no relative rate of phosphatidate formation was substantially higher than that observed in the presence of palmitoyl-CoA alone, it is apparent that several mixtures containing C-18 fatty acids gave considerably lower reaction rates. However, the formation of DG

TABLE 5. *Formation of TG from various species of radioactive DGs*

DG substrate	% DG → TG Acyl-CoA Substrate			
	Palm.	Oleoyl	Palm. oleoyl	Palm. linoyl.
Palm.	37.4	35.4	35.9	35.5
Oleoyl	32.8	34.3	34.8	36.5
Palm. Oleoyl	35.7	33.3	38.7	35.2
Palm. Linoyl	33.4	33.1	38.1	34.8

Palm., palmitoyl-CoA; Linoyl., linoleoyl-CoA.

from these various phosphatidates shows a preference for molecular species containing palmitate and a rather strong exclusion of phosphatidates containing palmitoleate. Although the latter was unusual for molecular species of phosphatidates, it does suggest that some specificity in this reaction may occur.

We next examined the specificity of DG acyltransferase. The various diglyceride species formed and utlized as substrates in these studies are shown in Table 4. A widely different pattern of diglyceride species were formed depending on the acyl-CoA substrates used as indicated earlier. The results shown in Table 5 indicate that these various species of diglycerides reacted with individual acyl-CoA derivatives or various mixtures in a similar manner to form triglyceride (TG). For each of the DG species used as substrate the rate of conversion to TG was nearly identical for each of the acyl-CoA mixtures studied. These results would indicate that the DG pool formed from GPA is freely utilized in the formation of TG and with no specificity except that determined by the acyl-CoA distribution in liver.

DISCUSSION

These studies have described the use of a new technique for the preparation of microsomal-bound lipid substrates in an investigation of the specificity of several reactions in liver glycerolipid biosynthesis. The use of these substrates avoids the inherent difficulties associated with aqueous dispersions of lipid substrates. Despite incubation of the microsomes with pure acyl-CoA derivatives under conditions unfavorable to the formation of acyl-CoA compounds *de novo,* a mixture of molecular species of the various lipid products was observed. It is concluded that the microsomal preparations contained small but significant amounts of acyl-CoA derivatives, which were incorporated under the usual incubation conditions. Alternative explanations for this observation exist, but it is obvious that a single molecular species of phosphatidate could not be formed by this technique.

The substantial specificity in phosphatidate formation observed previously was confirmed by this work (1–4). The predominant presence of saturated fatty acids in position 1 and unsaturated fatty acid in position 2 was noted and conformed to the molecular distribution observed in phosphatidate isolated from rat liver.

Initial findings indicated no major specificity in the formation of DG from phosphatidate using common acyl-CoA substrates in accordance with earlier reports (6, 8). However, further studies in which unusual fatty acyl derivatives were used in various mixtures indicated poor utilization of phosphatidates containing palmitoleate by phosphatidate phosphohydrolase. Thus minor specificity in this reaction probably exists, although the physiologic significance must be considered uncertain.

The studies of DG acyltransferase indicate no specificity for the diglyceride species used. However, only DG species formed from sn-GPA could be studied by these techniques, and the result does not exclude specificity for the considerably different molecular species of DG formed from lecithin by reversal of the choline phosphotransferase reaction (14).

The formation of lecithin from DG does show relative exclusion of highly saturated DG species. Although the latter must be considered a minor component of the DG pool, the results conform to the distribution of fatty acids found in rat liver lecithins. The relatively higher proportion of diunsaturated fatty acids in lecithins in contrast to the DG substrate also suggests preferential selection of these DG in the choline phosphotransferase reaction. Again these results are compatible with structural analysis of the lecithin and DG content of liver (15). The difference in the molecular species of DG and lecithin has been attributed to several factors, including a cycle involving prompt deacylation of newly formed lecithins with subsequent reacylation by more unsaturated fatty acids. The present findings suggest that specificity in the choline phosphotransferase reaction itself, as measured by these techniques, may contribute to the unique molecular species found in liver lecithins.

ACKNOWLEDGMENTS

The authors acknowledge the support of U.S. Public Health Service Grants AM09000 and AM18067 and the Netherlands Organization for Pure Scientific Research.

REFERENCES

1. Wilgram, G. F., and Kennedy, E. P. (1963): *J. Biol. Chem.*, 238:2615.
2. Smith, S. W., Weiss, S. B., and Kennedy, E. P. (1957): *J. Biol. Chem.*, 223:915.
3. Lamb, R. G., and Fallon, H. J. (1974): *Biochim. Biophys. Acta*, 348:166.
4. Eibl, H., Hill, E. E., and Lands, W. E. M. (1969): *Eur. J. Biochem.*, 9:250.

5. Lamb, R. G., and Fallon, H. J. (1970): *J. Biol. Chem.*, 245:3075.
6. Possmayer, F., Scherphof, G. L., Dubblemam, T. M. A. R., van Golde, L. M. G., and van Deenen, L. L. M. (1969): *Biochim. Biophys. Acta,* 176:95.
7. Yamashita, S., and Numa, S. (1972): *Eur. J. Biochem.,* 31:565.
8. Åkesson, B., Elovson, J., and Arvidson, G. (1970): *Biochim. Biophys. Acta,* 210:15.
9. Åkesson, B. (1969): *Eur. J. Biochem.,* 9:406.
10. van den Bosch, H., and Vagelos, P. R. (1970): *Biochim. Biophys. Acta,* 218:233.
11. Lee, T. C., and Snyder, F. (1973): *Biochim. Biophys. Acta,* 291:71.
12. Mudd, J. B., van Golde, L. M. G., and van Deenen, L. L. M. (1969): *Biochim. Biophys. Acta,* 176:547.
13. Kanoh, H. (1970): *Biochim. Biophys. Acta,* 218:249.
14. Kanoh, H. (1969): *Biochim. Biophys. Acta,* 176:756.
15. Hanahan, D. J. (1960): *Lipid Chem.,* p. 71. Wiley, New York.

Lipids, Vol. 1, edited by R. Paoletti, G. Porcellati, and G. Jacini. Raven Press, New York © 1976.

Metabolism of Phosphoglycerides and their Molecular Species in Brain

Giuseppe Porcellati and Luciano Binaglia

Istituto di Chimica Biologica, Università di Perugia, Facoltà Medica, C.P. n.3, 06100 Perugia, Italy

Studies carried out with mammalian tissues, chiefly liver (1, 2), have shown that glycero-3-phosphate (GPA) is acylated by acyl-CoA esters to yield phosphatidic acid (PA), which is the key intermediate in the subsequent reactions involved in lipid syntheses. PA may be synthesized, however, at least by two different additional pathways, i.e. by the phosphorylation of diacylglycerol (DG), as first reported by Hokin and Hokin in nervous tissue (3) and lately by others (4, 5), or by the acylation of dihydroxyacetonphosphate (6). PA produces different DG, which result in the immediate precursor used for *de novo* synthesis of diacyl-*sn*-glycero-3-phosphorylcholine or phosphatidylcholine (PC) and of diacyl-*sn*-glycero-3-phosphorylethanolamine or phosphatidylethanolamine (PE).

The composition of the molecular species of PC and PE is known to be regulated at many levels in the biosynthetic sequence, namely during the formation of PA and its subsequent transformation into DG by dephosphorylation, during DG utilization for phosphoglyceride synthesis, and by the acyl-exchange reactions involving position-specific enzymes and lysophospholipid intermediates (7). To these, another mechanism based on the reversal of cytidine diphosphate choline (CDP-choline) and CDP-ethanolamine diacylglycerol phosphotransferases (8–10) and on the activity of phosphatidylinositol phosphodiesterase (11) may be added.

Several studies (12–19), carried out with liver tissue, have distinguished different metabolic rates for the synthesis of the molecular species of membrane glycerolipids. Up to now, however, little or no information exists concerning whether the selection mechanisms delineated in liver are applicable to other tissues and, for our interest, to brain.

Since PC and PE are the major diacylglycerophosphatides of brain membranes, and their immediate primary precursor is DG, a study of the *in vivo* incorporation of 2-[^{3}H]glycerol into rat brain lipids and lipid precursors was undertaken. O'Brien and Geison (20) have recently reported some data related to this problem, but have not examined very short intervals from isotope administration and did not relate therefore completely the incorporation rates of glycerol into PA to those into DG, PC, and PE at the level of molecular subspecies. This report shows preliminary results about biosynthesis *in vivo* of these lipids at short time intervals (from 5 sec to 20 min) after intracerebral injection of 2-[³H]glycerol.

MATERIALS AND METHODS

Chemicals

Solvents were all glass-distilled and contained 0.1% (w/v) of 2,6-di-*tert*-butyl-*p*-cresol, as antioxidant. Ethyl ether was freed from peroxides, and chloroform was stabilized with 1% methyl alcohol. Phospholipase C from *Clostridium welchii* was supplied by Sigma Chemical Co. (St. Louis, Mo.) and pancreatic lipase by Nutritional Biochemical Corp. (Cleveland, Ohio). 2-[³H]glycerol, at specific activity (SA) of 500 mCi/mmole, was obtained from The Radiochemical Centre (Amersham, England). All chemical reagents were analytic grade products.

Injection and Removal of Brain Tissue

2-[³H]glycerol (100 μCl, 5 μl) was injected intracerebrally, in isotonic KCl solution, into the right parietal cortex of Sprague-Dawley male rats (120 to 140 g body weight) with a Hamilton microsyringe. The animals, which had been lightly anesthesized intraperitoneally with thiopentone, were prepared in a stereotaxic frame. Injection (1 sec) was made into 3 mm of cerebral surface. After predetermined intervals, the injected cerebral zone, together with the surrounding material, was quickly aspirated by means of a water-driven vacuum machine into a Potter-Elvehjem glass-tube that had been carefully precooled in liquid nitrogen. The same container was used for subsequent operations. The brain material, which was normally obtained in 1 sec of time, was about 10 to 20% of total brain. At longer time intervals, when diffusion of the injected material was much higher (150 sec onward), a larger amount of brain tissue was taken, in order to recover more thoroughly the labeled material. It is known, on the other hand, that the intracerebral route allows a relatively ready access to all membranous structures in brain cells.

Lipid Extraction

The tissue lipids were extracted in the cold as described by Kates (21). The extracts were washed twice with methanol:water (1:1, v/v) containing 1% (w/v) glycerol and 2.5% (w/v) NaCl and then with methanol:water (1:1, v/v).

Labeling of Lipid Classes

The washed lipid extract was analyzed for labeling the polar lipid classes by the two-dimensional thin-layer chromatography (TLC) of Horrocks (22), which separates the individual lipid classes, and in addition allows the plasmalogen subclass of a given lipid to be separated from the mixture of the alkyl, acyl, and diacyl derivatives.

A portion of the total extract was analyzed for the less polar lipids, after conversion of PA into the dimethyl-PA with diazomethane (12), by TLC on silica gel G plates with diethyl ether-petroleum ether (40 to 60°)-acetic acid (80:20:1, v/v/v) as the solvent (12). Triglycerides (TGs) were better resolved following Åkesson's procedure (23) on silica gel G plates using light petroleum (40 to 60°)-diethyl ether (90:10, v/v) as the developing solvent.

β-Monoglycerides were identified by TLC using authentic β-monostearin, prepared from distearin by incubation with pancreatic lipase as a reference standard. Two different chromatographic analyses were performed on TLC plates of silica gel H containing 5% boric acid:(a), one-dimensional TLC with diethyl ether-light petroleum ether (40 to 60°)-acetic acid (80:20:1-v/v/v); (b), bidimensional TLC with the same solvent as in (a) and then with chloroform-acetone (96:4, v/v).

The radioactive zones of the TLC plates were scraped into liquid scintillation counting vials, and the radioactivity content was determined.

Preparative Chromatography of Lipids

The preparation of PC and PE was carried out using the classical TLC procedure with chloroform-methanol-water (65:25:4, v/v/v), as the solvent. DG and dimethyl-PA were prepared by chromatography of total lipid extract following diazomethane treatment (23).

PC and PE were eluted from the prepared TLC plates by successive extractions with chloroform-methanol-acetic-acid-water (50:39:1:10, v/v/v/v), as described elsewhere (13). DG were eluted with distilled diethyl ether and dimethyl-PA with chloroform-methanol (2:1, v/v).

Preparation and Separation of Lipid Molecular Subspecies

The fractionation of lipid classes into molecular subspecies was achieved by argentation TLC on silica gel H (Merck, Darmstadt, West Germany) plates. DG were fractionated according to Åkesson et al. (14). PC was transformed into its corresponding diacylglycerol by phospholipase C treatment (24), and the resulting DG was examined as described previously. PE was fractionated in the form of N-acetyl-O-methyl derivatives, according to the method of Sundler et al. (25). Dimethyl-PA was fractionated according to Åkesson (12).

The TLC zones that contained lipid material were visualized with 2.7-dichloro fluorescein spray, and successive observations were made under UV light. The material was scraped from plates into test tubes, eluted with chloroform-methanol (2:1, v/v), and the extracts washed with 1 volume of methanol-water (1:1, v/v), containing 5% NaCl. The chloroform phases were transferred into liquid scintillation counting vials, evaporated nearly to dryness, and the radioactivity content determined, as further explained. In some experiments the chloroform phases were methanolyzed, and radioactivity was measured for both fatty acyl methyl esters (FAME) and glycerol-containing moieties.

Analytical Methods

The results of the argentation TLC procedure were examined by analysis of the fatty acid (FA) content of each spot. For this purpose the FA were analyzed by gas-liquid chromatography (GLC) of their FAME (14) in a Carlo Erba (Milan, Italy) Fractovap Model GV GLC instrument at 195°C. The stationary phase was 20% EGA on sylanized Chromosorb-P (100–200 mesh) in 150 × 0.3 cm stainless steel columns. The response of the flame ionization detector was quantitated by the triangulation method and each peak area was divided by the respective molecular weight to obtain the percent composition in moles of FA for each subspecies. By adding to the methanolysis mixture a 17:0 specimen, as an internal standard, the percent molar distribution of molecular subspecies for a given lipid class was obtained.

Data of percent molar distribution have also been obtained for PC and PE subspecies by determination of the phosphate content according to Ernster et al. (26).

Radioactivity was measured in a Packard Tri-Carb liquid scintillation spectrometer Model 3320, using Instagel Packard (Des Plaines, Ill.) as the scintillation solution. The calculation was carried out by adopting the channels ratio method.

RESULTS

Molecular Subspecies of PC, PE, DG and PA

Table 1 indicates the values of the pools for molecular subspecies of PC, PE, DG and PA, determined on the basis of the degree of unsaturation. The data

TABLE 1. *Pool sizes of rat brain PC, PE, PA, and diacylglycerols*[a]

Fraction	PC %	PC µmoles/brain	PE %	PE µmoles/brain	PA %	PA µmoles/brain	DG %	DG µmoles/brain
Saturated	20.2	4.24	12.0	1.35	9.5	0.11	2.7	0.003
Monoenoic	33.7	7.08	11.0	1.24	14.2	0.17	5.1	0.006
Dienoic	0.52	0.11	5.5	0.62	13.7	0.16	5.0	0.006
Trienoic	1.16	0.24	2.6	0.29	4.5	0.05	0.9	0.001
Tetraenoic	32.5	6.82	41.9	4.73	13.2	0.16	71.2	0.085
Pentaenoic	3.5	0.73	7.3	0.82	10.1	0.12	14.2	0.017
Hexaenoic	8.4	1.76	19.4	2.19	34.8	0.41		

[a] Mean values of four experiments each.

are given as mole% and as μmoles/brain, and have been used for calculation of SA in the subsequent experiments.

A very high percentage of tetraenes (chiefly 38 of carbon atom number with a smaller percentage of 36) is present in DG, which is much less evident in PC and PE and even less in PA. On the contrary, saturated and monoenoic species, which represent only about 8% of the DG, form more than 50% PC. A noticeable difference in species distribution exists between PA and DG, reflected in a lower content of tetraenes in the former and a lower content of dienes and hexaenes in the latter. Differences between PA and DG species have been reported to occur in brain tissue also by Baker and Thompson (27). Satisfactory similarity is seen between our data on PC subspecies and those of O'Brien and Geison (20), if exception is made of our higher value of tetraene species.

The differences reported in Table 1 are of interest in view of the possible different metabolic rates for different molecular species, and can imply different metabolic routes for molecules having various degrees of unsaturation.

Incorporation of Glycerol into Lipid Classes

The labeling of brain glycerolipids at short time intervals from glycerol injection (5 to 60 sec) is shown in Fig. 1. The amount of isotope rises first in a phosphorus-free lipid compound exhibiting a peak at 20 sec, whereas the labeling of the other lipids rises continually.

The phosphorus-free compound has been identified as β-monoglyceride by the following experiments: (a) the R_f value of the compound was found to be identical to that of authentic β-monostearin in the two TLC systems reported in Materials and Methods; (b) the chromatographic mobility in diethyl ether-petroleum ether

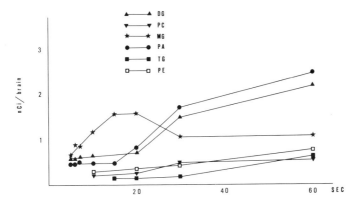

FIG. 1. Incorporation of intracerebrally injected [³H]glycerol into different brain lipids at short time intervals after administration. The data are means of six experiments; values for 60 sec are means of three experiments. PA, phosphatidic acid; DG, diacylglycerol; MG, monoglyceride; PE, phosphatidylethanolamine; PC, phosphatidylcholine; TG, triglyceride.

(40 to 60°)-acetic acid (80:20:1, v/v/v), as solvent did not change after diazome-thane treatment, thus confirming the absence of acidic groups in the molecule; (c) after 1 hr of acetylation of the isolated compound at 60°C in anhydrous pyridine-acetic anhydride (1:3, v/v), evaporation of the reagents, and TLC on silica gel G plates with chloroform-acetone (96:4, v/v), as the developing solvent, 60% of the original radioactivity migrated together with authentic pure 2-acyl-1,3-diacetyl glycerol and 40% in the position of β-acyl-αacetyl glycerol; (d) the FA pattern of the isolated compound was very similar to that of the monoglyc-eride (MG) isolated from mouse brain (28).

The labeling in DG and PA takes place at 15 to 20 sec after injection, although with lower radioactivity content than in MG. The labeling in the first two lipids then rises continually, whereas there is a drop in that of the MG. After about 25 to 30 sec after administration, PC and PE begin to be labeled significantly. TG begins to be labeled at 15 to 30 sec and then radioactivity rises.

It is possible that a direct esterification of glycerol may take place at the first time intervals with production of MG (although this reaction has not been reported to occur in nervous tissue). The MG might then contribute to DG production at the first time intervals, and then to PA through the kinase reaction.

Figure 2 shows that the SA of MG always exceeded for the time examined that of DG, although not with a distinct precursor-product relationship. Interest-ingly, within 5 to 100 sec or more, DG always had a higher SA than PA. It is

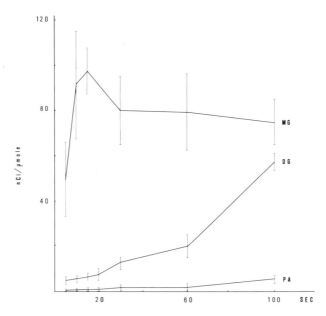

FIG. 2. Specific activities of brain **MG,** diacylglycerol **(DG),** and **PA** at short time intervals after [³H]glycerol administration. The data are means ± SD of six experiments; values for 60 and 100 sec are means of three experiments. For abbreviations, see Fig. 1.

necessary in order to reconcile these findings with the known mechanisms of lipid synthesis either to postulate that in *de novo* synthesis DG is derived from a pool or species of PA (microsomal PA?) with a SA much greater than that of the average for the total brain PA pool or to admit a different origin of the DG during the time considered.

Isotope uptake into various lipid classes at higher time intervals is reported in Fig. 3. Contrary to liver (14), no peaks for DG and PA labeling were observed at the times shown, and this finding is in line with the reports of Baker and Thompson (27), carried out in rat brain *in vivo* with glycerol, which indicated peaks of maximal SA at about 20 min and 30 to 60 min for PA, and for PC and PE, respectively. The difference may be caused by the way isotope is administered and not by the different ability to utilize the precursor. It is worth mentioning, in this connection, that GPA reaches a very high value of labeling already by approximately 15 sec after glycerol administration, as estimated according to the method of Hill et al. (16). Incorporation of radioactivity into GPA was already of the order of 2% at this time, whereas the labeling rate increased very slowly thereafter. This slow increase of labeling probably is reflected in an active utilization of glycerol by the glycerol kinase reaction and a rapid equilibration of GPA both with its precursors and products.

Figure 3 shows a continuous rise in the radioactivity content of phosphatidylinositol (PI), PE, PC, TG, and MG without a definite peak of maximun labeling. The radioactivity of TG rises continually. It is not known whether this lipid is formed in brain only *via* PA or also *via* MG as in other animal tissues (29–31), although the decreasing order of SA of MG, DG, and TG at these times should

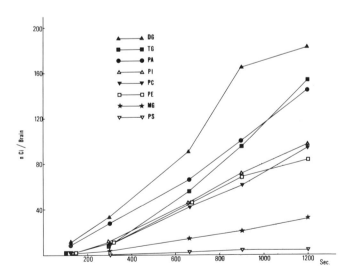

FIG. 3. The incorporation of intracerebrally injected [³H]glycerol into different brain lipids at long time intervals after administration. The data are means of three experiments each. **PI,** phosphatidylinositol; **PS,** phosphatidylserine. For other abbreviations see Fig. 1.

not deny this second hypothesis. The small radioactivity content of phosphatidyl-serine (PS), which begins to rise only after 300 to 400 sec after glycerol injection, may be accomplished by the activity of the base-exchange enzymic system successfully demonstrated in brain tissue *in vitro* (32). The radioactivity incorporated into PC and PE was almost completely found in their diacyl-subclasses, the labeling of plasmalogens representing only about 5% of the total radioactivity of each class. Only a very small labeling was observed in free FA at 20 min after injection.

The glycerol incorporation into the molecular subspecies of PA, DG, PC, and PE has been examined between 5 sec and 20 min after administration. The data in Tables 2 and 3, expressed as a percentage of radioactivity distribution among molecular subspecies of a given lipid class, indicate in general that the percent distribution in the subspecies is not constant with time even between 5 and 60 sec, and therefore that the distribution of radioactivity at the earlier times is very different from those at longer times. Particularly there is a gradual accumulation of saturated species of PA and a rather similar phenomenon for those of DG; the monoene fraction of both PA and DG shows an initial increase of their percent radioactivity and a noticeable decrease thereafter. Probably the results at the very early stages of labeling could be related to an early metabolic event, which was mentioned before, namely the production of MG. This phenomenon may have taken place very quickly, but is difficult to be interpreted and probably of small quantitative value.

Table 2 shows also a gradual increase of glycerol incorporation into pentaene and hexaene species of PA, which may be reflected in the successive increase in same species of DG. It is not excluded that the increase in the polyunsaturated DG at these times may be due to transacylation reactions with monoene DG, which show a corresponding decrease in their percentages of labeling in this period (Table 2). After 300 sec the percentage of distribution in DG and PA remains almost constant if exception is made for the increase in saturated DG and PA and for that in the hexaene species of PA.

No striking differences among PC subspecies (Table 3) are seen, if exception is made for the high initial conversion of glycerol into monoenoic species that can be related to the highly radioactive pool of monoenoic PA (Table 2). At times higher than 10 min, a decreasing percentage of labeling is seen in the order $\Delta 1 \rightarrow \Delta 6 \rightarrow \Delta 0 \rightarrow \Delta 4 \rightarrow \Delta 2 \rightarrow \Delta 5$. Owing to the large pool of saturated and monoenoic PC in rat brain (Table 1), these last data may be interpreted to indicate a higher SA of the polyunsatured species of PC compared to that of $\Delta 0$ or $\Delta 1$ species at 20 min, a finding that is comparable to the SA values published recently by O'Brien and Geison (20) for periods of 30 min after glycerol administration.

The very high initial uptake of radioactivity into saturated PE (Table 3) is worth mentioning; it may be related to an initial preferential utilization of saturated DG (formed *via* MG?). This preferential utilization of saturated species is not evident thereafter. The very high uptake of glycerol into hexaene PE begins to take place at approximately 30 to 50 sec after administration, increases sharply

TABLE 2. Incorporation of radioactivity (% of each lipid class) into molecular subspecies of PA and diacylglycerol (DG) of rat brain at different time intervals after the intracerebral injection of [³H]glycerol[a]

Time (sec)	PA						DG					
	Δ0	Δ1	Δ2-3	Δ4	Δ5	Δ6	Δ0	Δ1	Δ2	Δ3	Δ4	Δ5-6
5	31.5	62.5	0	2.5	1.5	2	75	20	0	0	0	5
10	30.5	57.5	0	7.5	2	2.5	70	22.5	0.2	0	0.5	6.5
20	33	57.5	0.5	12.5	7	7	54	21.7	0.2	0	2	15
30	35	28	0.5	10	7	19	50.5	29.5	0.5	0	3.5	16.5
60	38	17	0.4	4	18	23	52.5	7.5	1.3	0	7	32
150	42.5	20	0.5	2.0	14	21	55.5	8	2	0	7	24.5
300	49.5	22.5	0.5	1.2	8	17.5	62.5	6.5	2.5	1	11	18
600	57.0	16	0	2.5	9.5	13.5	65	6.5	2	1	8.5	15.5
900	64	9	0	2.5	7.5	18.2	73	6.5	1.7	1.2	6	12.5
1200	67	2	0	0.2	1	31.5	74	6.0	2	1.2	5.2	17.5

[a] Mean values of four experiments each.

TABLE 3. Incorporation of radioactivity (% of each lipid class) into molecular subspecies of PC and PE of rat brain at different time intervals after intracerebral injection of [³H]glycerol[a]

Time (sec)	PC						PE				
	Δ0	Δ1	Δ2–3	Δ4	Δ5	Δ6	Δ0	Δ1	Δ2–3	Δ4	Δ5–6
20	12	63	9	7.5	2	7.5	70.5	7	8.5	4	10
30	14.5	51	9.7	9.5	2.3	13	39	16	13	12	20
60	13	40.5	9.7	19	2.5	16	8	22.5	19	22	28.5
150	16.5	37.5	8.5	9.5	7	21.5	6.2	8	11	16.3	58.5
300	17.5	25.5	8.5	9.5	4.5	36.5	7	5	6	11	71
600	21	32	7	14	5.5	24	1	10	5.5	10	73.5
900	25	30.5	6	13	5.2	22	1.5	5	5.5	12	76
1200	31	29	5	12	5.3	21	2.2	0.3	5.5	14	78

[a] Mean values of four experiments each.

thereafter, and reaches values of about 75% at 20 min. It is possible at this time that an active transacylation mechanism or an active utilization of polyunsaturated DG formed by degradation of endogenous phospholipids are at play in explaining these results.

DISCUSSION

All together the results of the present work indicate a certain selectivity of glycerol incorporation into brain lipids. Even at the earliest time, the incorporation of glycerol into PA and DG occurred considerably in the saturated and monoenoic fractions. Owing to the relatively low content of these species in the total PA and DG pool in rat brain (Table 1), these considerations seem valid also in terms of SA. Probably a mixed intervention due to selectivity and pool compartmentalization (namely inhomogeneity of the pools) may occur in this phenomenon.

The high level of incorporation of glycerol into saturated DG in brain may probably be accounted for by a high conversion to TG, although this has not been proved in our experiments. More than 80% of the rat brain TG fatty acids are saturated or monounsaturated (11). The metabolic inertness of saturated DG to be converted into PC and PE is found also in studies with liver tissue (14) but is not absolute.

Metabolic sources for brain DG other than via PA have to be taken into account in order to explain our results (Figs. 1 and 2) and the striking differences in molecular species between brain DG and PA (Table 1). The close relationships between the subspecies of PA and DG, reported for rat liver (14), may exist in brain only for a few of them. The intervention of MG in the sequential scheme of reactions leading to PA in brain remains an open and interesting question. The reactions leading to lyso-PA and PA from MG and DG have been amply demonstrated in brain, but no information is as yet available on MG conversion to DG nor on esterification of glycerol to MG.

The work of Johnston et al. (33) has shown that DG formed from MG is not utilized in intestine for PC synthesis contrary to those formed from GPA. This has not been ascertained, however, for brain. The possible participation, on the other hand, of this scheme of reaction in phospholipid metabolism in brain may contribute for the higher SA observed for DG, as compared to PA, in the saturated species (Fig. 4). Either a very active pool of PA may explain these data or this PA must come from MG *via* DG.

The preferential utilization of monoenoic DG for PC synthesis (Fig. 4), as compared to PE, indirectly confirms our previous results (34), which showed that rat brain microsomes are more able to utilize *in vitro* cytidine diphosphocholine (CDP-choline) than cytidine diphosphoethanolamine (CDP-ethanolamine) in the presence of monoenoic DG.

No preferential utilization seems to exist for tetraene DG for PC and PE synthesis, as reported in Fig. 5, whereas a much better utilization of highly

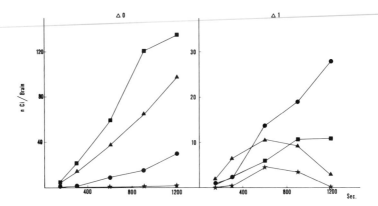

FIG. 4. Conversion rates (percentage of incorporated radioactivity for each lipid class) of saturated **(left)** and monoenoic **(right)** diacylglycerol (DG) (■—■—■), PA (▲—▲—▲), PC (●—●—●), and PE (★—★—★) synthesis, after intracerebral injection of [³H]glycerol. The data of the figure may be converted into SA values by examining the pools of the molecular species of the single lipid classes (Table 1).

unsaturated DG takes place in brain for PE synthesis, as compared to PC (Fig. 5). Transacylation reactions may explain, however, these data, although previous work *in vitro* (34) has demonstrated that rat brain microsomes are more able to utilize CDP-ethanolamine than CDP-choline in the presence of polyunsaturated DG.

It is very difficult to draw conclusions at the present time from the results reported in this work. It is difficult also to calculate from the present experimental data turnover rates for single molecular species, because rather strong variations

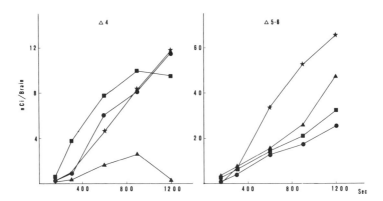

FIG. 5. Conversion rates (percentage of incorporated radioactivity for each lipid class) of tetra-enoic **(left)** and pentahexaenoic **(right)** diacylglycerol (DG) (■—■—■), PA (▲—▲—▲), PC (●—●—●), and PE (★—★—★) synthesis, after intracerebral injection of [³H]glycerol. The data of the figure may be converted into SA values by examining the pools of the molecular species of the single lipid classes (Table 1).

with time occurred in calculating the values of rate constants (12, 14, 35). This could have been caused by either alternate utilization of reaction products or compartimentalization among cell types or within the cell. Even with these limitations, the dienoic and tetraenoic species of DG are converted better than other DGs to PC at short time intervals (20 to 90 sec), whereas hexaenoic species are preferred in these time intervals for the synthesis of PE. At longer intervals the PC species with polyunsaturated FA are synthesized more rapidly than the monoenoic and saturated species.

To what extent these results are caused by substrate specificity of the enzymes involved is not known. In our opinion, the selective utilization of DG may be caused by both structural compartimentalization and substrate specificity, although previous work *in vitro* (34) has indicated a certain degree of specificity in the various steps of phospholipid synthesis.

SUMMARY

The biosynthesis of rat brain PC and PE has been studied *in vivo* at short time intervals after the intracerebral injection of 2-[^{3}H]-glycerol. PA appears to be synthesized by two different pathways: the glycerophosphate pathway and a MG pathway. The MG, which has been identified by means of its chemical and chromatographic properties and its fatty acid pattern, is the most highly labeled lipid at time intervals as short as 15 sec from administering. At that time glycerophosphate also is synthesized in large amounts. The specific activity of DG is higher than of PA for the first 20 min from administration. Monoenoic DGs are preferred for PC synthesis as compared to PE; tetraenoic DGs are similarly incorporated into PC and PE, whereas hexaenoic DGs are preferred for PE synthesis.

ACKNOWLEDGMENTS

This investigation was aided by a research grant from the Consiglio Nazionale delle Ricerche, Rome (contract No. 73.00730.04). Dr. R. Roberti has collaborated at some phases of the work. Skillful technical assistance has been provided by Mr. P. Caligiana.

REFERENCES

1. Kornberg, A., and Pricer, W. E., Jr. (1953): *J. Biol. Chem.*, 204:345.
2. Kennedy, E. P. (1953): *J. Biol. Chem.*, 201:399.
3. Hokin, M. R., and Hokin, L. E. (1959): *J. Biol. Chem.*, 234:1381.
4. Lapetina, E. G., and Hawthorne, J. N. (1971): *Biochem. J.*, 122:171.
5. Friedel, R. O., and Schanberg, S. M. (1971): *J. Neurochem.*, 18:2191.
6. Hajra, A. K. (1968): *Biochem. Biophys. Res. Commun.*, 33:929.
7. Lands, W. E. M., and Merkl, I. (1963): *J. Biol. Chem.*, 238:898.
8. Weiss, S. B., Smith, S. W., and Kennedy, E. P. (1958): *J. Biol. Chem.*, 231:53.
9. Kanoh, H., and Ohno, K. (1973): *Biochim. Biophys. Acta*, 326:17.

10. Kanoh, H., and Ohno, K. (1973): *Biochim. Biophys. Acta,* 306:203.
11. Keough, K. M. W., MacDonald, G., and Thompson, W. (1972): *Biochim. Biophys. Acta,* 270:337.
12. Åkesson, B. (1970): *Biochim. Biophys. Acta,* 218:57.
13. Arvidson, G. A. E. (1968): *Eur. J. Biochem.,* 5:415.
14. Åkesson, B., Elovson, J., and Arvidson, G. (1970): *Biochim. Biophys. Acta,* 210:15.
15. Hill, E. E., Lands, W. E. M., and Slakey, S. P. M. (1968): *Lipids,* 3:411.
16. Hill, E. E., Husbands, D. R., and Lands, W. E. M. (1968): *J. Biol. Chem.,* 243:4440.
17. Parkes, J. G., and Thompson, W. (1973): *Biochim. Biophys. Acta,* 306:403.
18. Kanoh, H. (1970): *Biochim. Biophys. Acta,* 218:249.
19. Trewella, M. A., and Collins, F. D. (1973): *Biochim. Biophys. Acta,* 296:51.
20. O'Brien, J. F., and Geison, R. L. (1974): *J. Lipid Res.,* 15:44.
21. Kates, M. (1972): In: *Laboratory Techniques in Biochemistry and Molecular Biology, Vol. 3,* p. 349. North-Holland, Amsterdam.
22. Horrocks, L. A. (1968): *J. Lipid Res.,* 9:469.
23. Åkesson, B. (1969): *Eur. J. Biochem.,* 9:463.
24. Lands, W. E. M., and Hart, P. (1966): *J. Am. Oil. Chem. Soc.,* 43:240.
25. Sundler, R., and Åkesson, B. (1973): *J. Chromatogr.,* 80:233.
26. Ernster, L., Zetterström, R., and Lindberg, O. (1950): *Acta Chem. Scand.,* 4:942.
27. Baker, R. R., and Thompson, W. (1972): *Biochim. Biophys. Acta,* 270:489.
28. Sun, G. Y. (1970): J. Neurochem., 17:445.
29. Senior, J. R., and Isselbacher, K. J. (1962): *J. Biol. Chem.,* 237:1454.
30. Johnston, J. M., and Brown, J. L. (1962); *Biochim. Biophys. Acta,* 59:500.
31. Rao, G. A., and Johnston, J. M. (1966): *Biochim. Biophys. Acta,* 125:465.
32. Porcellati, G., Arienti, G., Pirotta, M., and Giorgini, D. (1971): *J. Neurochem.,* 18:1395.
33. Johnston, J. M., Paultauf, F., Schiller, C. M., and Schultz, L. D. (1970): *Biochim. Biophys. Acta,* 218:124.
34. Porcellati, G. (1972): *Adv. Enzyme Regul.,* 10:83.
35. Zilversmit, D. B., Entenman, C., and Fishler, M. C. (1943): *J. Gen. Physiol.,* 26:325.

Lipids, Vol. 1, edited by R. Paoletti, G. Porcellati, and G. Jacini. Raven Press, New York © 1976.

Metabolism of Retina Acylglycerides and Arachidonic Acid

Nicolás G. Bazán, Marta I. Aveldaño, Haydee E. Pascual de Bazán, and Norma M. Giusto

Instituto de Investigaciones Bioquímicas, Universidad Nacional del Sur, Bahía Blanca, Argentina

Since excitable membrane polar lipids may require readily available sources of precursors for their biosynthesis and cell functioning, an important aspect of lipid neurochemistry is the detailed knowledge of acyl group distribution and regulation at the level of lipid metabolic intermediates such as acylglycerides and free fatty acids (FFAs). This chapter summarizes recent studies from our laboratory showing that in the retina, although diacylglycerols and triacylglycerols are metabolites of the *de novo* biosynthetic pathway of lipids, they may also be involved in other lipid metabolic reactions. Data are also presented showing the high metabolic rate of the free arachidonate pool. Additional experiments indicating several aspects of the *de novo* biosynthesis of glycerolipids are discussed. We have used the isolated retina since it is an integral part of the central nervous system and because it can be easily obtained and incubated *in vitro* without harmed surfaces as is the case with brain slices (1).

EXPERIMENTAL

The incubation of retinas was performed at 37°C in the medium described by Ames and Baird-Hastings (2); unless otherwise specified all the media contained 2 mg/ml of glucose and were gassed with $O_2:CO_2$ (95:5). In experiments regarding the release of FFAs, different lipid-free bovine serum albumin concentrations were included in the media. After incubation lipids were extracted as described elsewhere from tissue (3) and medium (4), respectively.

The experiments relative to arachidonic acid incorporation were performed by incubating cattle retinas for 1 hr in the presence of 5 μCi of [5,6,11,12,14,15-^{3}H]arachidonic acid (SA 8 Ci/mmole) per incubation flask. Each flask contained one retina. The retinas were thoroughly washed with fresh medium and incubated for an additional period of 20 min in labeled fatty acid-free medium (5). The incorporation of glycerol was followed by incubating cattle and toad retinas during different time intervals in the presence of 5 μCi of U-[^{14}C]glycerol per incubation flask (SA 7.4 mCi/mmole), each containing one cattle and twenty toad retinas, respectively (6, 7).

In all cases the incubations were stopped by homogenization of the retinas in chloroform/methanol. The total lipid extracts were alternatively filtered to obtain the lipid-free dry weights or centrifuged to perform a residual lipid extraction (5). In the latter case, the protein content was determined. The phospholipids and neutral lipids were separated by gradient-thickness (3, 9) and two-dimensional (6, 7) thin-layer chromatography (TLC). Diacyl- and triacylglycerols were obtained from unincubated retinas and their fatty acid (FA) composition was determined by gas-liquid chromatography (GLC) (8). Free arachidonic acid and triacylglycerol arachidonate were quantitated by GLC using methyl nonadecanoate as the internal standard. Radioactivity was measured as previously described (5–7) in a Packard Tricarb liquid scintillation counter.

ACYL GROUP DISTRIBUTION IN RETINA DIACYLGLYCEROLS AND TRIACYLGLYCEROLS

The content of diacylglycerols greatly varies when comparing retinas of different vertebrates. On a lipid-free dry weight basis, the highest content is found in the toad retina where a concentration approximately fourfold higher than in cattle and rabbit is achieved (10). The acyl group distribution also markedly differs (Fig. 1). Docosahexaenoate comprises approximately 42% of toad retina diacylglycerol FAs. Thus more than 80% of these diglycerides (DGs) are composed of hexaenoic molecular species, whereas only about 8% docosahexaenoate is present in the cattle diacylglycerols. In the cattle retina on the contrary, an enrichment in arachidonate was found, which together with stearate conform about half of the fatty acids, suggesting that the 1-stearoyl-2-arachidonoyl-*sn*-glycerol is the dominant molecular species as is the case in rat brain (22). Moreover, since the content of diacylglycerol arachidonate is similar in both species (10), a high proportion of the difference is attributable to the presence of the hexaenoic molecular species in the toad retina diacylglycerols.

In the toad retina, the apolar side-chain composition of total phospholipids is very similar to that of diacylglycerols. In cattle, however, contrasting differences were observed. Owing to the arachidonate and stearate content, an apparent

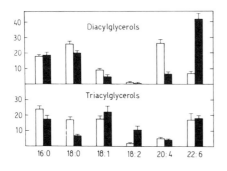

FIG. 1. Comparative acyl group distribution of cattle *(open bars)* and toad *(black bars)* retina diacylglycerides **(upper)** and triacylglycerides **(lower).**

relationship of a portion of the DG pool with phosphatidylinositol (PI) may exist (10).

The total content of triacylglycerols is also higher in the toad than in the cattle retina (10) without such profound differences in the proportions of arachidonate and docosahexaenoate as in diacylglycerols (Fig. 1). A direct precursor-product relationship between most of the diacyl- and triacylglycerols is not apparent when comparing the composition of both neutral lipids in each animal species.

FREE ARACHIDONIC ACID POOL OF RETINA

Since one of the features of neural cell membranes is the large content of highly unsaturated acyl groups in their lipids, we have searched for the free unsaturated pool sizes and metabolism. The neural tissue FFAs may be rapidly modified by different functional states. The level of endogenous FFAs is the result of the activity of several enzymes, mainly those involved in the acylation and deacylation of complex lipids. Free arachidonic acid, as well as other polyenoic fatty acids is rapidly increased in the mammalian central nervous system at the onset of ischemia (3, 13, 14), following electroshock (3, 15, 16), and by drug-induced convulsions (16). We have proposed that they may arise from the activation of phospholipases A_2 (3, 13–16). Since the total polar lipids of neural tissues including the retina are enriched in docosahexaenoate in relation to arachidonate and in the free fatty pool an inverse situation is seen, a very high metabolic activity of arachidonate-containing lipids is evidenced in the retina in agreement with studies carried out in rat brain (17, 18).

The incubation of retinas in the medium described by Ames and Baird-Hastings (2), even in the presence of glucose, gives rise to an enlargement of the FFA pool that is more noticeable when the tissue:medium volume ratios are lower. It was suggested that they may be produced by a relative hypoxia in this condition, since a severalfold increase in FFAs occurs in retinas incubated in anoxia (19, also *unpublished work*). However, in the presence of albumin, a significant proportion of FFAs is lost from the tissue and can be recovered in the medium. Figure 2C shows that free arachidonic acid increases about twofold between 5 and 20 min of incubation and that this change may be entirely ascribed to the extracellular FA, since a nearly constant level is maintained in the tissue. The measurement of triacylglycerol arachidonate under these conditions clearly demonstrates that the FFAs do not originate in these neutral lipids (Fig. 2C) as is also the case in anoxia.

The incubation of retinas under pure nitrogen without glucose gives rise to a marked increase in free arachidonic acid that accumulates in the tissue (Fig. 2B). The presence of albumin in the incubation medium promotes the outflow of a great proportion of the produced acid. If the retina is incubated under normal O_2 tension in the presence of increasing albumin concentrations, redistribution of free arachidonic acid takes place between the tissue and the medium without any apparent effect of albumin on the total free arachidonate content (Fig. 2A).

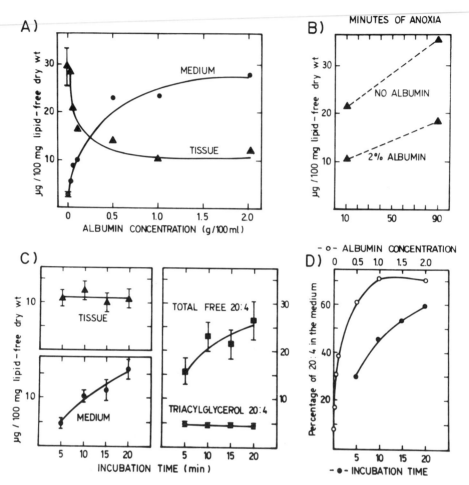

FIG. 2. Modifications of the kinetics involved in the maintenance of cattle retina free arachidonic acid *in vitro.* **(A)** Incubations were made in the presence of glucose (2 mg/ml) for 20 min at 37°C under $O_2:CO_2(95:5)$. Each point is one retina incubated in 7 ml of medium (●) The mean values ± SD from four samples incubated without albumin are also included (▲). **(B)** Incubations were performed for 90 min in the same conditions without glucose and under pure N_2 in the presence of 2% albumin **(lower)** or without albumin **(upper).** Only tissue arachidonic acid concentration is presented. **(C)** Two retinas per sample were incubated as in A during the specified times in the presence of 0.5% albumin. The mean values ± SD from four samples are indicated. The arachidonate content of triacylglycerol 20:4 **(lower)** and total free 20:4 **(upper)** is also shown. **(D)** The percentage of medium arachidonic acid was obtained from the experiments presented in A and C. ○, albumin concentration; ●, incubation time.

When analyzing the percentage of the FA transferred to the medium on increased albumin concentrations, a hyperbolic curve is obtained (Fig. 2D). Further increases of albumin do not produce an additional decrease of tissue arachidonate, suggesting that the medium-free acid may be originated in cytoplasmic peripheral membranes or in a plasma membrane-related pool. In the FFA translocation mechanisms, a membrane acylation-deacylation cycle may be involved in addition to a possible energy-independent passive diffusion process. The latter cannot be excluded, since in complete anoxia the FFA outflow is not inhibited; however, on long incubations the integrity of the retina may be damaged. The acylation-deacylation cycle may be an important component of the FFA uptake mechanisms. Free arachidonic acid is easily released from the retina in the presence of albumin, in addition an effective intake occurs since the radioactive FA not only appears in the FFA pool but also a high rate of esterification was observed (5).

[3H]ARACHIDONIC ACID LABELING OF RETINA LIPIDS

Figure 3 presents the distribution of [3H]arachidonic acid in retina lipids after 1 hr of incubation followed by a subsequent incubation for an additional 20 min in labeled FA-free medium. Surprisingly, in spite of the low arachidonate content of retina triacylglycerols, a noticeably high percentage of the incorporated FA is present in this lipid. Moreover, although the tetraenoic FA is in trace amounts when compared with the total polar lipids, the incorporation in triacylglycerols is more than half the amount present in the former (Fig. 2) and represents the highest incorporation when compared with any of the individual phospholipid classes. Among the latter the maximum incorporation rate is attained by PI closely followed by phosphatidylcholine (PC).

DE NOVO BIOSYNTHESIS OF ACYLGLYCERIDES BY [14C]GLYCEROL

The *de novo* biosynthesis of retina lipids was surveyed by following the incorporation of [14C]glycerol during *in vitro* incubations. Only 18 hr after intracerebral injection of [14C]glycerol, recycling significantly contributes to lipid labeling (20). In the retina incubated with this labeled precursor, lack of recycling is suggested due to the labeling sequence and by the finding of radioactivity in the appropriate moiety of lipids. In Fig. 4 a comparison is presented of the labeling kinetics by [14C]glycerol of toad and cattle retina lipids. In both a sequence phosphatidic acid (PA)-diacylglycerol-triacylglycerol is apparent. In the cattle retina the label-

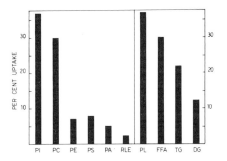

FIG. 3. [3H]arachidonic acid incorporation into bovine retina lipids. PI, phosphatidylinositol; PC, phosphatidylcholine; PE, phosphatidylethanolamine; PS, phosphatidylserine; PA, phosphatidic acid; RLE, residual lipid extract; PL, total phospholipids; FFA, free fatty acids; TG and DG triacyl- and diacylglycerols, respectively.

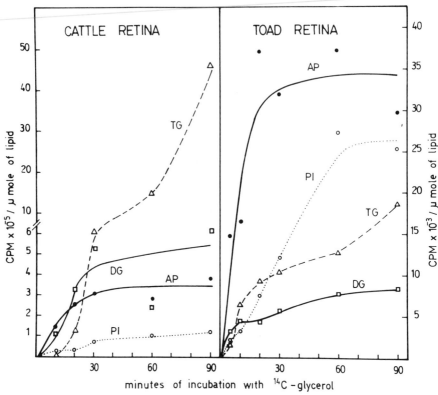

FIG. 4. Changes in specific activities of cattle **(left)** and toad **(right)** retina lipids as a function of incubation time in the presence of [^{14}C]glycerol. Abbreviations as in Fig. 3.

ing of PC and phosphatidylethanolamine (PE), the major polar lipids, is lower than in PI. Although in the toad retina a similar labeling is observed, a faster and higher uptake of the labeled precursor occurs in PI (Table 1). A similar incorporation profile was found in the toad brain and retina several hours after

TABLE 1. *Comparative labeling with* [^{14}C]*glycerol of retina lipids*

Retina lipid	Cattle		Toad	
	SA (× 10^{-5})	Percent uptake	SA (× 10^{-3})	Percent uptake
PA	3.89	3.7	27.6	8.1
PI	1.26	8.7	23.5	28.4
Diacylglycerols	2.28	3.8	8.2	10.3
Triacylglycerols	45.97	60.0	19.8	12.1
PC	0.26	17.5	1.5	27.5
PE	0.11	6.0	1.0	11.9
Phosphatidylserine	0.02	0.4	1.0	1.8

In both instances the results were obtained after 90 min of incubation. SA, specific activity as in Fig. 4.

in vivo injection of the precursor, indicating that the incubation conditions employed did not modify the metabolic course of glycerol in the toad retina (7). Furthermore this implies that the biosynthesis of PI in the unstimulated poikilothermic neural tissues proceeds at high rates. After 90 min of incubation, a predominance of the pathway toward triacylglycerols was evidenced in the cattle retina (Fig. 3 and Table 1).

CONCLUSIONS

Marked compositional and metabolic differences between cattle and toad retina acylglycerides were found. In the latter approximately 80% of the total diacylglycerols is made up of molecular species containing docosahexaenoate, possibly 1-palmitoyl-2-docosahexaenoyl-*sn*-glycerol and 1-stearoyl-2-docosahexaenoyl-*sn*-glycerol. The toad retina triacylglycerol composition shows, on the other hand, less than 20% docosahexaenoate besides other dissimilarities with diacylglycerols. From the [14C]glycerol labeling in *in vitro* experiments, it is concluded that the retina contains an efficient glycerol kinase activity (6, 7) and that in the *de novo* biosynthesis of toad retina lipids the following sequence is operating: PA-diacylglycerols-triacylglycerols. Thus, a great proportion of the docosahexaenoate-containing DG pool may not be involved in the formation of triglycerides (TGs). Phospholipids are richly endowed with hexaenoic acid in the toad retina (10) and in the rod outer segments of the photoreceptors isolated from the retinas of several animal species (for references see 10). The docosahexaenoate-containing diacylglycerols may result from a metabolic specialization of the toad retina to maintain available adequate concentrations to be used when the cytidine diphosphate-phosphocholine (and ethanolamine) phosphotransferases are required to synthesize phospholipids. Such a step should be tightly regulated in the retina since very active membranogenesis takes place in the photoreceptor cell supporting the continuous rod outer segment renewal stimulable by light exposure and temperature (21). In the retina, as shown here in the toad, a selective acylation step may be present in addition to introduce the hexaenoic acyl group in specific pools preceeding PA biosynthesis from α-glycerophosphate. When the amount of hexaenoic DGs exceeds the required amount, a new acylation step may yield TGs. The role, if any, of phospholipases C and of the "back reaction" (22) in the formation of the hexaenoic molecular species of diacylglycerols remains to be evaluated.

The *de novo* biosynthetic pathway in the unstimulated toad retina *in vivo* and *in vitro* seems to be actively committed to the biosynthesis of PI through PA (7). In the isolated synaptosomes, [14C]glycerol labeling of PI also occurs rapidly; however, this result was not interpreted as reflecting biosynthesis (23).

A completely different labeling outline was evidenced when cattle retinas were incubated under identical conditions than those of the toad in the presence of [14C]glycerol. Although in the cattle retina the same sequence was followed by the precursor as in the toad, a rapid formation of high specific activity TGs was the feature (6). Moreover a relatively lower synthesis of PI takes place. The high rate of triacylglycerol biosynthesis may reflect the presence of a branch regulatory point. As mentioned above, the phosphotransferases may show high specificity toward DGs according to the cellular site and requirements. In previous studies using brain, similar suggestions were made (24, 25). Among the factors involved in the activity of the enzymes that may shift the *de novo* lipid biosynthesis that should be borne in mind is the relative concentration in the tissue of ions (26–28) and substrates (28) as shown in nonneural tissues.

The cattle retina acylglycerides also showed other prominent peculiarities. Although TGs contain less than 5% arachidonate, the uptake of [³H]arachidonic acid amounts to about a third of the total recovered as acyl group in retina lipids (Table 2).

Unlike triacylglycerols at least half of the DGs are made of the tetraenoic molecular species and incorporate 18% of the labeled FA. More than half of the tritiated arachidonic acid is esterified to phospholipids (Table 2), PI and PC accounting for the largest proportion (Fig. 3). The incorporation of [¹⁴C]glycerol, on the other hand, showed 63% of labeling in TGs, in a comparative experiment, 34% in polar lipids, and only 3% in diacylglycerols (Table 2). This indicates that although a slow *de novo* biosynthesis of PI and mainly of PC (Table 1) occurs in the cattle retina under the conditions used, a very high [³H]arachidonic acid uptake occurs (Fig. 3). Thus in both phospholipids an active acylation-deacylation cycle may be operating yielding tetraenoic molecular species.

Because triacylglycerol arachidonate does not change in conditions that lead to an increase of free arachidonic acid such as in anoxia and a high proportion of the supplied acid is incorporated into this fraction, a metabolic role of triacylglycerides distinct from the supply of metabolic energy is suggested. In addition to the observation of a high metabolic rate in brain TGs (29, 30), the high specific activity of TG arachidonate may indicate their involvement as acyl group donors in certain biosynthetic routes of polar lipids and as conforming to a specific pool leading through a lipase to prostaglandin biogenesis (5). Moreover the elevated levels of free arachidonic acid accumulated in the tissue or released to the medium in the presence of an available acceptor, suggests that it originates

TABLE 2. *Labeling of cattle retina lipids by different precursors[a]*

Lipids	[³H]arachidonic acid (% incorporated)[b]	[¹⁴C]glycerol (% incorporated)
Diacylglycerols	18	3
Triacylglycerols	31	63
Total phospholipids	52	34

[a] The data for each isotope were obtained in separate experiments using cattle retinas incubated as described in the text for 60 and 90 min in the presence of [³H]arachidonic acid and [¹⁴C]glycerol, respectively, and then incubated in a radioactivity-free medium for a further 20-min period.

[b] The percentage for [³H]arachidonate was calculated excluding the radioactivity present in the FFA pool.

in membrane phospholipids by the interplay activity of deacylating and reacylating enzymes. This mechanism may be important in the exchange of FA between the tissue and the extracellular fluids.

ACKNOWLEDGMENTS

This work was supported by the Consejo Nacional de Investigaciones Científicas y Técnicas, Argentina. M.I.A. is a Research Fellow of this Institution. Thanks are owed to Dr. Eduardo Charreau from the Instituto de Biología y Mediciná Experimental for kindly allowing us to use the liquid scintillation counter.

REFERENCES

1. Ames, A., III, and Gurian, B. S. (1960): *J. Neurophysiol.,* 23:676.
2. Ames, A., III, and Baird-Hastings, A. (1956): *J. Neurophysiol.,* 19:201.
3. Bazán, N. G. (1970): *Biochim. Biophys. Acta,* 218:1.
4. Aveldaño, M. I., and Bazán, N. G. (1974): *FEBS Lett.,* 40:53.
5. de Bazán, H. E. P., and Bazán, N. G., *submitted for publication.*
6. Giusto, N. M., and Bazán, N. G., *submitted for publication.*
7. de Bazán, H. E. P., and Bazán, N. G., *submitted for publication.*
8. Aveldaño, M. I., and Bazán, N. G. (1973): *Biochim. Biophys. Acta,* 296:1.
9. Bazán, N. G., and Bazán, H. E. P. (1975): In: *Research Methods in Neurochemistry,* edited by N. Marks and R. Rodnight, pp. 309-324. Plenum Press, New York.
10. Aveldaño, M. I., and Bazán, N. G. (1974): *J. Neurochem.,* 23:1127.
11. Aveldaño, M. I., and Bazán, N. G. (1972): *Biochem. Biophys. Res. Commun.,* 48:689.
12. Keough, K. M. W., McDonald, G., and Thompson, W. (1972): *Biochim. Biophys. Acta,* 270:337.
13. Bazán, N. G., de Bazán, H. E. P., Kennedy, W. P., and Joel, C. D. (1971): *J. Neurochem.,* 18:1387.
14. Aveldaño, M. I., and Bazán, N. G., *Brain Res., in press.*
15. Bazán, N. G., and Rakowski, H. (1970): *Life Sci.,* 9:501.
16. Bazán, N. G. (1971): *J. Neurochem.,* 18:1379.
17. Baker, R. R., and Thompson, W. (1972): *Biochim. Biophys. Acta,* 270:489.
18. Yau, T. M., and Sun, G. Y. (1974): *J. Neurochem.,* 23:99.
19. Giusto, N. M., and Bazán, N. G. (1973): *Biochem. Biophys. Res. Commun.,* 55:515.
20. Benjamin, J. A., and McKhann, G. M. (1973): *J. Neurochem.,* 20:1111.
21. Young, R. W. (1967): *J. Cell Biol.,* 33:61.
22. Kanoh, H., and Ohno, K., *This Volume.*
23. Lunt, G. G., and Lapetina, E. G. (1970): *Brain Res.,* 17:167.
24. McCaman, R. E., and Cook, K. (1966): *J. Biol. Chem.,* 241:3390.
25. O'Brien, J. F., and Geison, R. L. (1974): *J. Lipid Res.,* 15:44.
26. Liberti, J. P., and Jezyk, P. F. (1970): *Biochim. Biophys. Acta,* 210:221.
27. Jamdar, S. C., and Fallon, H. J. (1973): *J. Lipid Res.,* 14:509.
28. Erbland, J. F., Brossard, M., and Marinetti, G. V. (1967): *Biochim. Biophys., Acta,* 137:23.
29. Sun, G. Y., and Horrocks, L. (1971): *J. Neurochem.,* 18:1963.
30. Yavin, E., and Menkes, J. H. (1973): *J. Neurochem.,* 21:901.

Lipids, Vol. 1, edited by R. Paoletti, G. Porcellati, and G. Jacini. Raven Press, New York © 1976.

The Mechanism of Action of Pancreatic Phospholipase A

A. J. Slotboom, W. A. Pieterson, J. J. Volwerk, and G. H. de Haas

Laboratory of Biochemistry, State University, Transitorium 3, Padualaan 8, Utrecht, The Netherlands

Our main interest in the mechanism of action of lipolytic enzymes is based on their very high activity toward substrates that form aggregated structures in water, such as monolayers, micelles, liposomes, or emulsions. Why are these enzymes almost unable to degrade substrates in molecularly dispersed form, whereas the same substrate is very effectively hydrolyzed after incorporation in a lipid-water interface? A possibility, advocated recently by Wells (1) to explain the activating effect of micellar aggregates on the action of snake venom phospholipase A, might be the much lower entropy of activation for hydrolysis of aggregated substrate molecules. Aggregation reduces rotational and translational energy of the monomers and might force the essential groups of the substrate molecule to occupy the surface of the micelle.

Another mechanism for interfacial activation of lipolytic enzymes, proposed many years ago by Desnuelle et al. (2), implies a reversible adsorption of the enzyme to the interface followed by a conformational change in the active site of the protein induced by the interface. Highly efficient interaction between protein and substrate molecules could occur as the latter diffuse laterally in the interface.

In recent years a large number of pure phospholipases A_2 have been isolated from different sources such as mammalian pancreas, snake venom, and bee venom, their amino acid composition has been reported (3–5). The enzymes turned out to be low molecular weight single-chain proteins that owe their high stability to the presence of many disulfide cross-links. The primary structure has been determined for the porcine pancreatic (3), bee venom (4), and *Bitis gabonica* venom enzyme (5); however, information concerning the amino acid residues involved in substrate binding and catalysis remains scanty. Detailed kinetic analysis has been reported only on the *Crotalus adamanteus* phospholipase A_2 (6, 7) and on the porcine pancreatic enzyme (8). There is evidence that both enzymes, although catalyzing the same reaction, do so by different mechanism.

Pancreatic phospholipase A_2 has been isolated in pure form in our laboratory from pig, ox, horse, and sheep pancreas and from all these sources two isoenzymes were obtained, differing slightly in amino acid composition. Figure 1 gives the amino acid sequence of the porcine enzyme.

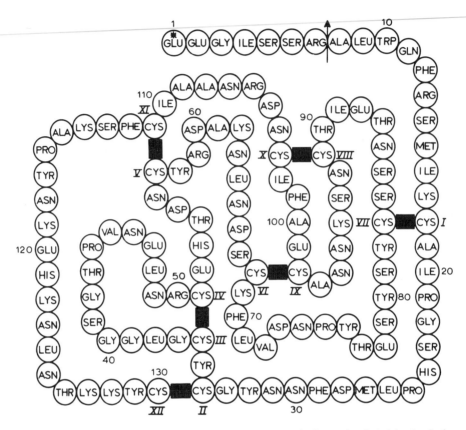

FIG. 1. Primary structure of porcine pancreatic (pro)phospholipase A₂. Asterisk stands for pyroglutamic acid.

The amino acid chain consists of 123 residues starting *N*-terminally with alanine-8 (Ala-8) and having ½ cystine-130 (Cys-130) as the *C*-terminal amino acid.[1] The pancreatic enzymes are all produced by the pancreas in an enzymically inactive zymogen form, containing an additional heptapeptide at the *N*-terminal site. For the different animals, the amino acid composition of the activation peptide is not the same; however, activation of the zymogens into the active phospholipase by trypsin always occurs by a specific cleavage of the arginine-7 (Arg-7)·Ala-8 linkage. Upon release of the activation peptide, a conformational change in the enzyme can be observed by fluorescence spectroscopy (9). The single tryptophan (Trp) in position 10, located so close to the newly formed *N*-terminal Ala-8 allows the hypothesis that the α-NH$_3^+$ group moves to a hydrophobic environment and probably forms an ion pair with an acidic group buried

---

[1] The numbering of the amino acid residues throughout the paper starts with the *N*-terminal amino acid of the zymogen.

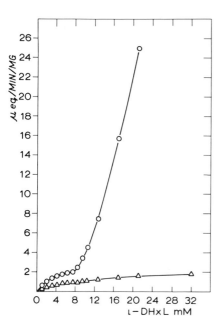

FIG. 2. Michaelis curves showing the activity of phospholipase A_2 ($\bigcirc$) and its zymogen ($\triangle$) as a function of dihexanoyl-lecithin concentration. Assay conditions: 0.5 mM NaAc; 0.1 M NaCl; pH 6.0; 40°C.

in a more apolar part of the molecule. The importance of this conformational isomerization for the interfacial activity of phospholipase A is discussed later in this chapter.

The pancreatic enzymes are characterized by an absolute requirement for Ca^{2+}, and no other metal ion can substitute for it (10). This suggests a specific role of Ca^{2+} in the catalytic mechanism of the enzyme. Ultraviolet difference spectroscopy and equilibrium gel filtration demonstrated that a single Ca^{2+} ion is bound to the enzyme close to the active site residue histidine-53 (His-53) (10, 11). Spectroscopic titration using fluorescence and high-resolution nuclear magnetoresonance (nmr) confirmed the role of His-53 in Ca^{2+} binding and showed a considerable drop in pK value upon interaction with the metal ion *(unpublished results)*. The zymogen was shown to possess a similar affinity for Ca^{2+} and binds this metal ion in a 1 : 1 ratio, again close to His-53. Apparently both proteins are characterized by an identical metal ion-binding site. In addition it could be demonstrated by a number of techniques that both the active phospholipases and their zymogens interact with monomeric substrate molecules even in the absence of Ca^{2+}. Also this ligand is bound to both proteins in the close neighborhood of His-53 (11). Therefore it has been concluded that the active site of phospholipase A preexists already in the zymogen and that the zymogen should display enzyme activity (12); this is an agreement with the recent report on proteolytic zymogens (13, 14). Figure 2 shows the catalytic power of phospholipase A and its zymogen using a synthetic water-soluble substrate, dihexanoyl-L-α lecithin.[2]

---

[2] 1,2-dihexanoyl-*sn*-glycero-3-phosphorylcholine.

   This short-chain phospholipid is characterized by a critical micelle concentration (CMC) of approximately 9.5 mM. It is clear that active phospholipase A
hydrolyzes this lecithin in a normal way at substrate concentrations where only
monomers are present. The activity, however, is very low, and only after passing
the CMC is there a tremendous increase in activity toward the micellar aggregates. The zymogen is also able to catalyze the hydrolysis of dihexanoyl-lecithin
monomers with a V_{max} value of about 50% of that of the active enzyme proving
that the catalytic site is fairly similar in both proteins. Apparently the most
important difference between both proteins is the the presence of an interfacial
recognition site (IRS) on the active phospholipase A, whereas the zymogen lacks
this site and does not interact with a lipid-water interface. Most probably this
IRS will be of a hydrophobic character and must be located in a position topo-
graphically distinct from the active site. This could be shown by switching off
the catalytic site of the enzyme through a specific irreversible blocking of His-53
using an active site-directed haloketone (11, 12). Such a protein completely lost
its catalytic power; however, it could be shown to possess a fully identical affinity
for lipid-water interfaces.
   Quantitative measurements of the affinity constant between phospholipase A
and micellar lipid-water interfaces can be effected using equilibrium gel filtration
on Sephadex. Scatchard plots of such binding experiments done at different
equilibrating concentrations yield affinity constants and the dependence of the
latter on pH can be studied. Figure 3, in which the affinity constants between
phospholipase A and a micellar lipid-water interface are plotted as a function
of pH according to Dixon (15), shows two straight lines, having slopes of 0 and
−1 intersecting at pH 8.1. Because the neutral lecithin molecule has a net charge
of zero in the pH range studied, the observed change reflects a difference of one
proton between two ionic forms of the enzyme. Therefore it has to be concluded
that the action of the IRS is controlled by a single amino acid residue having
a pK close to 8.1. This amino acid has been shown to be the N-terminal alanine
of the chain (9, 12).
   Figure 4 compiles in a schematic way our hypothesis of the activation mechanism and the formation of the IRS in the active enzyme upon zymogen

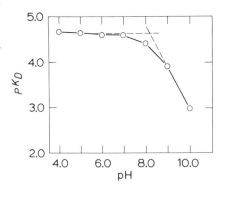

FIG. 3. Dixon-plot of the effect of pH on the interaction between phospholipase A_2 and the mixed micellar lipid system consisting of equimolar amounts of D-didecanoyl-lecithin + 1-myristoyl-L-lysolecithin.

PRECURSOR   PHOSPHOLIPASE A

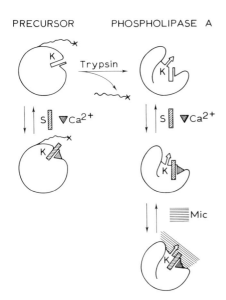

FIG. 4. Schematic representation of pancreatic phospholipase A₂ and its zymogen, showing the monomer-binding site (∏) and calcium-binding site (∨), which together constitute the catalytic region (K). During tryptic activation the zymogen loses its activation heptapeptide (∿X) and undergoes a conformational change in which an ion pair is formed between the α-NH₃⁺ group of the N-terminal alanine and a buried carboxylate. Linked to the formation of this salt bridge is the creation of the recognition site (⇑) for interfaces.

activation. Probably the formation of an ion pair between the newly formed α-NH$_3$$^+$ group of the N-terminal alanine and a buried carboxylate triggers not only the induction of the IRS but at the same time a small rearrangement of active-site residues allowing a much more effective catalysis.

Where is the IRS located and how is it constructed? It seems evident that the architecture of the IRS must be different for the various lipolytic enzymes and probably determined by the *in vivo* function of these enzymes. For the pancreatic enzymes that attack in the intestine the loosely constructed mixed micelles of phospholipids and bile salts, one can expect that a weak penetrating power might be sufficient to allow interaction with the substrates avoiding at the same time an uncontrolled breakdown of surrounding membrane systems. The fact that the pancreatic phospholipase A has indeed a weak penetration site for organized lipid-water interfaces has been demonstrated using a series of synthetic lecithins of different chain length. Although the dihexanoyl-, diheptanoyl-, and dioctanoyl-lecithin forming micellar systems in water are readily attacked by the pancreatic enzyme without addition of detergents, the higher homologues form bilayer structures and the dense packing of the lecithin molecules in the liposome makes them inaccessible to the enzyme (16). A more quantitative study of the interactions between phospholipase A and lecithins using monolayer techniques (17) gave similar results.

Recently, Dr. Op den Kamp in our laboratory has been able to demonstrate this property of porcine pancreatic phospholipase A in a different way (18). Although long-chain lecithin dispersions in water in general are not degraded by the pancreatic enzyme, he observed a rapid hydrolysis of the liposomes at the transition temperature of the lecithin (see Fig. 5, right).

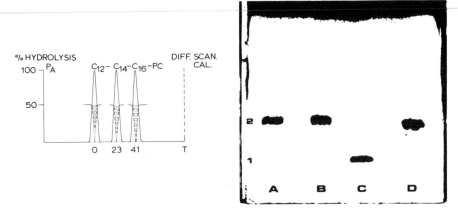

FIG. 5. Left: Differential scanning calorimetry (▼) and phospholipase A_2 hydrolysis of three synthetic lecithins containing two identical acyl chains: dilauroyl-lecithin (C_{12-}), dimyristoyl-lecithin (C_{14-}), and dipalmitoyl-lecithin (C_{16-}). **Right:** TLC of dimyristoyl-lecithin after treatment with phospholipase A from pig pancreas. Samples of 0.4 μmole of lecithin in the form of liposomes in Tris-HCl buffer pH 7.2, containing 1 mM $CaCl_2$, were incubated for 15 min. (A) Control experiment without phospholipase A_2. Incubation at 15° (B), 23° (C), and 37°C (D). **1** is lysolecithin; **2** is lecithin.

Both below and above the transition temperature, the highly ordered lipid molecules are not accessible to the pancreatic enzyme, and hydrolysis does not take place. Figure 5 (left) shows how the transition temperatures, as determined by differential scanning calorimetry, coincide with enzyme hydrolysis for dilauroyl-, dimyristoyl-, and dipalmitoyl-lecithin. Apparently, at the transition temperature, the melting process of the fatty acid chains induces local perturbations in the architecture of the bilayer allowing the enzyme to interact with the interface. The venom phospholipases that are characterized by a more powerful penetration site (17) do not show this behavior. Although these enzymes too are unable to interact with lecithin dispersions below the transition temperature, rapid enzymatic degradation is observed both at and above the transition temperature. These enzymes do interact with liposomes having their acyl chains in the ordered liquid state.

The fact that the formation of a salt-bridge between the newly formed N-terminal α-NH_3^+ group and a probably buried carboxylate triggers the induction of the interfacial recognition site, suggests that the apolar N-terminal region of the protein chain might be involved directly in the recognition process. In order to study the function of the N-terminal part of the enzyme, chemical modification and substitutions of amino acids in the N-terminal region have been effected. To avoid unwanted reactions of the ϵ-NH_2 groups of the nine lysine residues of the protein, a fully ϵ-amidinated phospholipase A has been prepared as shown in Fig. 6 using methylacetimidate.

From Table 1 it is clear that the fully amidinated phospholipase A (AMPA) still possesses 60% of the activity of the native enzyme, while its affinity for

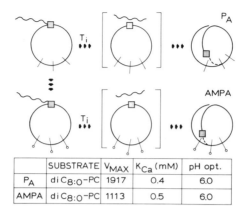

	SUBSTRATE	V_{MAX}	K_{Ca} (mM)	pH opt.
P_A	di $C_{8:0}$-PC	1917	0.4	6.0
AMPA	di $C_{8:0}$-PC	1113	0.5	6.0

FIG. 6. *(Upper circles)* Schematic presentation of the tryptic conversion of prophospholipase A into the active enzyme (P_A). /|\ stands for the nine ϵ-NH_2 groups present in both the zymogen and active enzyme. *(Lower circles)* Tryptic conversion of fully ϵ-amidinated prophospholipase A into AMPA. /|\ stands for the nine amidinated ϵ-NH_2 groups. Enzyme activities (in μmoles ·min^{-1}·mg·$protein^{-1}$) were determined using dioctanoyl-lecithin (di$C_{8:0}$-PC) as substrate, 0.1 M NaCl, pH 6.0, 40°C. □, N-terminal L-Ala.

calcium and pH optimum does not change. Edman degradation of AMPA produces Des-Ala-AMPA, and from Table 1 it is evident that removal of the N-terminal amino acid of phospholipase A does reduce the enzyme activity on monomeric substrates from 100 (AMPA) to about 10%. However, this modifica-

TABLE 1. *Influence of the N-terminal amino acid (□) on the kinetic parameters of AMPA*

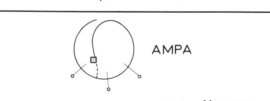

AMPA

	Micelles V_{max} (%)	Monomers V_{max} (%)	K_s(mM)
L-Ala	100	100	4.2
Gly	65	51	1.2
β-Ala	27	27	3.5
D-Ala	2.9	24	20
0(Des-Ala)	0.6	8.2	12
L-Ala-L-Ala	0.7	20	20
t. Boc-L-Ala	0.5	59	18
Pyroglu · · · Arg · Ala	0.3	47	18
Ala · Leu · Phe	37	51	4.2
L-Ala from Des-Ala	106	—	—

V_{max} values are compared using dihexanoyl-lecithin (monomers) and dioctanoyl-lecithin (micelles) as substrate. Assay: 0.1 M NaCl; pH 6.0; 40°C.

tion almost completely switches off the interfacial recognition site: lipid-water interfaces are not hydrolyzed anymore (100 → 0.6%). These changes in activity are not caused by damage to other regions of the protein chain during the Edman degradation as could be shown by reintroduction of a covalently bound L-Ala into Des-Ala-AMPA by means of *N-t*-butyloxycarbonyl-L-alanine *N*-hydroxy-succinimido ester. After removal of the *N*-blocking group, an AMPA was obtained in all respects identical to the original AMPA (enzyme activity on micellar substrates was found to be 106% of that of "native" AMPA).

Starting from Des-Ala-AMPA we have elongated the protein chain with a new amino acid different from the native L-Ala. Table 1 shows that substitution of the original L-Ala by Gly or β-Ala as *N*-terminal residue yields enzymes having not only monomer but also appreciable interfacial activity. Fluorescence spectroscopic titration of these modified proteins confirmed the presence of a salt bridge in these enzymes. Chopping off the first three amino acid residues Ala, Leu, and Trp by repeated Edman degradation and introduction of a covalently bound tripeptide Ala·Leu·Phe yielded a protein in which the native Trp-10 is replaced by a Phe residue. This modification leads to a phospholipase A possessing considerable monomer and interfacial activity.

On the other hand, it is evident from Table 1 that blocking of the α-NH$_2$ function of L-Ala in AMPA, chain elongation of AMPA itself with a L-Ala residue and even substitution of the native L-Ala by the stereoisomeric D-Ala gives rise to proteins with considerable monomer activity, indicating a functionally active catalytic site. However, these proteins appeared to be devoid of interfacial activity. They behave as the naturally occurring zymogen (Table 1, Pyroglu . . . Arg·AMPA) and fluorescence titration showed the absence of the salt bridge. The results of these studies leave little doubt that there is a close relationship between the existence of an ion-pair (α-NH$_3^+$ . . . $^-$OOC) and highly efficient interfacial enzyme activity. However, they do not prove that the apolar *N*-terminal part of the chain itself interacts with lipid-water interfaces. It cannot be precluded that a conformational change in the enzyme upon formation of the salt bridge induces the formation of the recognition site in another part of the molecule.

Ultraviolet difference, fluorescence, and nmr spectroscopic studies are underway using specifically labelled proteins and lipid micelles to detect the localization of this site. In addition, detailed knowledge about the three-dimensional structure of the enzyme will soon become available through high-resolution X-ray spectroscopy.

ACKNOWLEDGMENTS

We would like to express our indebtedness to our colleagues, past and present, who have collaborated with us on this project. The results presented in this chapter are an outcome of association with a number of investigators both inside and outside of our laboratory. Special thanks are due to Dr. S. Maroux, Prof. M. Lazdunski, Dr. R. Verger, Dr. J. P. Abita, and Dr. J. C. Vidal for their contributions.

REFERENCES

1. Wells, M. A. (1974): *Biochemistry,* 13:2248.
2. Desnuelle, P., Sarda, L., and Ailhaud, G. (1960): *Biochim. Biophys. Acta,* 37: 570.
3. de Haas, G. H., Slotboom, A. J., Bonsen, P. P. M., Nieuwenhuizen, W., van Deenen, L. L. M., Maroux, S., Dlouha, V., and Desnuelle, P. (1970): *Biochim. Biophys. Acta,* 221:54.
4. Shipolini, R. A., Callewaert, G. L., Cottrell, R. C., Doonan, S., and Vernon, C. A. (1971): *Eur. J. Biochem.,* 20:459.
5. Botes, D. P., and Viljoen, C. C. (1974): *J. Biol. Chem.,* 249:3827.
6. Wells, M. A. (1972): *Biochemistry,* 11:1030.
7. Wells, M. A. (1974): *Biochemistry,* 13:2248.
8. de Haas, G. H., Bonsen, P. P. M., Pieterson, W. A., and van Deenen, L. L. M. (1971): *Biochim. Biophys. Acta,* 239:252.
9. Abita, J. P., Lazdunski, M., Bonsen, P. P. M., Pieterson, W. A., and de Haas, G. H. (1972): *Eur. J. Biochem.,* 30:37.
10. Pieterson, W. A., Volwerk, J. J., and de Haas, G. H. (1974): *Biochemistry,* 13:1439.
11. Volwerk, J. J., Pieterson, W. A., and de Haas, G. H. (1974): *Biochemistry,* 13:1446.
12. Pieterson, W. A., Vidal, J. C., Volwerk, J. J., and de Haas, G. H. (1974): *Biochemistry,* 13:1455.
13. Kassell, B., and Kay, J. (1973): *Science,* 180:1022.
14. Uren, J. R., and Neurath, H. (1974): *Biochemistry,* 13:3512.
15. Dixon, M. (1953): *Biochem. J.,* 55:161.
16. de Haas, G. H., Bonsen, P. P. M., Pieterson, W. A., and van Deenen, L. L. M. (1971): *Biochim. Biophys. Acta,* 239:252.
17. Verger, R., Mieras, M. C. E., and de Haas, G. H. (1973): *J. Biol. Chem.,* 248:4023.
18. Op den Kamp, J. A. F., de Gier, J., and van Deenen, L. L. M. (1974): *Biochim. Biophys. Acta,* 345:253.

Lipids, Vol. 1, edited by R. Paoletti, G. Porcellati, and G. Jacini. Raven Press, New York © 1976.

Lysophospholipases: Purification, Properties, and Subcellular Distribution

H. van den Bosch and J. G. N. de Jong

Laboratory of Biochemistry, State University of Utrecht, Transitorium 3, Padualaan 8, Utrecht, The Netherlands

The catabolism of phosphoglycerides in most biologic systems seems to proceed *via* the consecutive action of phospholipase A and lysophospholipases. During this process virtually no lytic lysophosphoglycerides accumulate, and the concentration of these compounds in membrane structures is commonly kept very low. Although reacylation of lysophosphoglycerides may play an important role in controlling their concentration, lysophospholipases (EC 3.1.1.5.) have to be considered in order to explain the turnover of the phosphate moiety of phosphoglycerides. In view of these considerations, it is not surprising that lysophospholipases have been found in almost any cell where they were searched for (for a review see ref. 1). Despite this widespread occurrence, only few attempts have been made to purify this type of lipolytic enzyme.

In recent years we have purified several enzymes with lysophospholipase activity to the point where specificity studies can be made. In this work we used a specific assay for lysophospholipase activity by measuring the release of radioactive fatty acid from synthetic 1-[1-^{14}C]palmitoyl lysolecithin.

We first purified a lysolecithin-hydrolyzing enzyme from beef pancreas (2). When studying the substrate specificity of this enzyme, we found that various other compounds, containing carboxylester bonds, such as short-chain triglycerides and *p*-nitrophenylacetate, were hydrolyzed at comparable or even much higher rates than was lysolecithin (3).

The results with the pancreatic enzyme provoked the question whether only pancreatic lysophospholipase is that nonspecific or whether this is a general property of lysophospholipases. This question is very important since we have evidence that the pancreatic enzyme is excreted. Thus it is quite conceivable that the broad substrate specificity of this enzyme would be related to other physiologic functions than participation in the catabolism of membrane phosphoglycerides.

We have therefore undertaken the purification of the lysophospholipase activity from beef liver. Interestingly, after solubilization of the lysophospholipase activity from this tissue by *n*-butanol treatment and chromatography on DEAE-Sephadex columns, two peaks with lysophospholipase activity emerged well separated from the column. We have provisionally denoted these enzymes lysophos-

TABLE 1. *Esterolytic properties of lysolecithin hydrolyzing enzymes*

Enzyme	SA (U/mg)		
	1-Acyl lysolecithin	Tributyrin	p-Nitrophenyl-acetate
Beef pancreas lysophospholipase	6.2	49	20
Beef liver lysophospholipase I	1.40	3.8	2.1
Beef liver lysophospholipase II	1.36	268	11

pholipase I and lysophospholipase II. They comprise about 25 and 75% of the total activity, respectively. Both enzymes were purified, 3,600-fold and 770-fold, respectively, to obtain homogeneous preparations (compare with Fig. 6). The molecular weights were estimated to be about 25,000 for lysophospholipase I and 60,000 for lysophospholipase II both by Sephadex G-100 filtration and SDS-disk gel electrophoresis. Both enzymes are optimally active at slightly alkaline pH values and neither requires Ca^{2+} or other bivalent metal ions for enzymic activity.

In view of the esterolytic properties found earlier for the pancreatic enzyme, the activity of the purified beef liver lysophospholipases versus tributyrin and p-nitrophenylacetate was investigated. Both beef liver enzymes exhibit esterolytic properties comparable to the pancreatic enzyme, although important variations can be noted for the different enzymes in the relative rates of hydrolysis of the investigated substrates (Table 1).

The results of these limited specificity studies gave rise to the question: can lysophospholipase activity be ascribed to a side action of nonspecific carboxylesterases? In this respect it is interesting to note that as yet no physiologic function has been assigned to these nonspecific carboxylesterases (4). This problem was investigated by assaying several highly purified carboxylesterases for lysophospholipase activity. The results of these assays (Table 2) clearly showed the high specific activity of carboxylesterases toward p-nitrophenylacetate and tributyrin. None of the esterases tested, however, showed any appreciable activity in hydrolyzing lysolecithin. Thus, it can be concluded that lysophospholipase activity

TABLE 2. *Assay of carboxylesterases for lysophospholipase activity*

Enzyme source	SA (U/mg)		
	1-Acyl lyso-lecithin	Tributyrin	p-Nitrophenyl-acetate
Pig liver[a]	0.00	79	63
Pig liver[b]	0.07	286	211
Beef liver[b]	0.00	24	182

[a] Commercial preparation.
[b] Kindly donated by Professor Dr. K. Krisch, Kiel, West Germany.

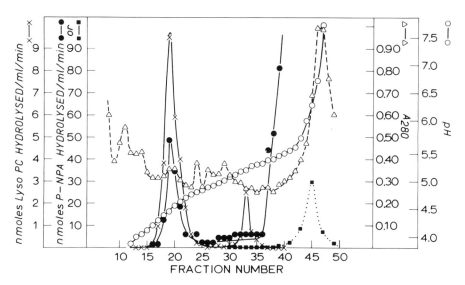

FIG. 1. Isoelectric focusing of lipid-free beef liver extract. Fractions of 2 ml were collected and assayed for lysophosphatidylcholine (lyso PC, x —— x) and *p*-nitrophenylacetate hydrolyzing activity (P-NPA in nmoles min⁻¹·ml⁻¹, ● —— ●; P-NPA in nmoles. min⁻¹·ml⁻¹ × 10⁻² (■ —— ■). pH (○ —— ○) and A$_{280}$ (△ —— △) are also shown. Cathode is at the right.

is not simply a side action of nonspecific carboxylesterases, although the lysophospholipase preparations isolated thus far certainly exhibit esterolytic properties.

We have attempted to show that these esterolytic properties are intrinsic to the lysophospholipases from beef liver. This was accomplished using the technique of isoelectric focusing. Figure 1 shows the results obtained when a crude butanol extract from beef liver was subjected to isoelectric focusing in an ampholine gradient from pH 4 to 6. As was expected the lysophospholipase activity gave two peaks with isoelectric points of pH 5.2 (lysophospholipase I) and of pH 4.5 (lysophospholipase II). The lysophospholipases were well separated from the main carboxylesterase peak, which banded well above pH 5.5. Coinciding with lysophospholipase II a second small peak of esterase activity versus *p*-nitrophenylacetate was detected, which apparently is not due to tailing off of the main carboxylesterase peak, as intermediate fractions contained zero esterase activity. The esterase activity of lysophospholipase I in the crude extract used in this experiment was too low to be detected.

However, when the 3,600-fold purified preparation of lysophospholipase I was isoelectrofocused on the same column, the esterase activity coincided with the lysophospholipase I activity (Fig. 2). Likewise the esterase activity in the 770-fold purified preparation of lysophospholipase II coincided with the lysophospholipase (Fig. 3). In both experiments no esterase activity was recovered in the region above pH 5.5, where a possible contaminating carboxylesterase was expected to peak (compare with Fig. 1). Thus the esterase activity versus *p*-nitrophenylace-

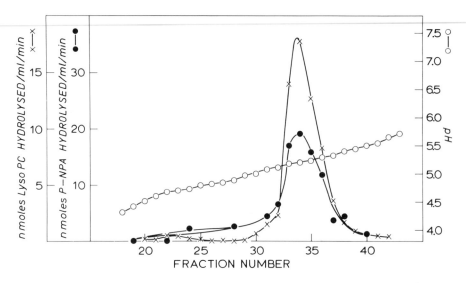

FIG. 2. Isoelectric focusing of a 3,600-fold purified preparation of lysophospholipase I. Symbols are as in Fig. 1.

tate is an intrinsic property of beef liver lysophospholipase. It may be good to recall here that many enzymes have been shown to possess the capacity of hydrolyzing *p*-nitrophenylacetate, not only hydrolases such as lipase (5), chymotrypsin (6), and lysozyme (7) but also glycero-3-phosphate dehydrogenase (8), aldehyde dehydrogenase (9) and carbonic anhydrase (10).

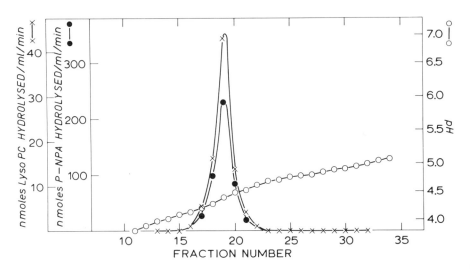

FIG. 3. Isoelectric focusing of a 770-fold purified preparation of lysophospholipase II. Symbols are as in Fig. 1.

TABLE 3. *Ratio of lysophospholipases in subcellular fractions from beef liver*

Subcellular fraction	Lysophospholipase I (%)	Lysophospholipase II (%)
Nuclei, debris	22	78
Mitochondria	38	62
Lysosomes	4	96
Microsomes	2	98
Cytosol	89	11

Recently we determined the subcellular localization of the two lysophospholipases from beef liver as part of a study to see if these enzymes are in any way related. In order to measure the contribution of each of the two enzymes to the total lysophospholipase activity in each subcellular fraction, the enzymes were solubilized and separated by DEAE-Sephadex chromatography for each subcellular fraction individually. The results of these experiments are compiled in Table 3. The microsomal fraction contained essentially only the lysophospholipase II, as does the lysosomal fraction. In contrast the supernatant fraction contained mainly the lysophospholipase I. The crude mitochondrial fraction contained both enzymes. Since we now know the ratio and recoveries of both lysophospholipases in each fraction, their subcellular distribution can be calculated. It is clear from Fig. 4 that lysophospholipase II follows a typical microsomal distribution. It is not present in the cytosol and from specific activity data of lysophospholipase II and various marker enzymes it could easily be concluded that the presence of lysophospholipase II in the crude lysosomal and mitochondrial fractions can be accounted for by microsomal contamination (data not shown).

Lysophospholipase I showed a bimodal distribution (Fig. 4) with highest relative specific activity in the cytosol and in mitochondria. This enzyme is virtually absent from the lysosomal and microsomal fraction. Thus, unlike the lysophospholipase II in the crude mitochondrial fraction, which is caused by microsomal contamination, the lysophospholipase I appears to be intrinsic to beef liver mito-

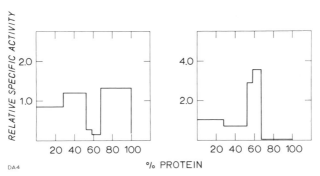

FIG. 4. Calculated relative specific activities of lysophospholipase I **(left)** and lysophospholipase II **(right)** versus percentage of total recovered protein in subcellular fractions from beef liver.

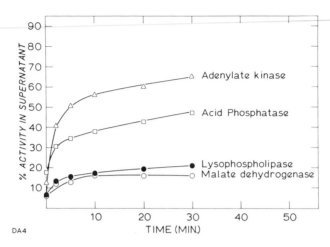

FIG. 5. Release of enzymic activities during hypotonic treatment of purified beef liver mito-chondria.

chondria. In accord with this hypothesis, a further purification of the crude mitochondria over sucrose gradients, effecting a decrease in the microsomal contamination, resulted in an almost complete removal of lysophospholipase II, with retention of lysophospholipase I. When purified mitochondria were dialyzed against distilled water the lysophospholipase I was recovered exclusively in the soluble fraction, suggesting that the enzyme originated from either of the soluble compartments of mitochondria. Upon incubation of purified mitochondria in hypotonic medium, lysophospholipase I was released simultaneously with malate dehydrogenase (Fig. 5), indicating that lysophospholipase I from mitochondria is present in the same compartment that accommodates malate dehydrogenase, i.e., the matrix fraction.

At present it is not known whether the lysophospholipase I from mitochondria is completely identical with the cytosolic enzyme, although no indications for differences were obtained during the purification procedures.

The subcellular distribution suggested that lysophospholipase I and II are completely different enzymes. Theoretically, however, the cytosol lysophospholipase I (MW 25,000) could be a soluble fragment of the microsomal lysophospholipase II (MW 60,000). Such a fragment could presumably be formed by proteolytic enzymes during the homogenization and centrifugation procedures. To check this possibility, homogenates were incubated for various times at room temperature and then used to determine the ratio of lysophospholipase I and II. It can be seen from Table 4 that ageing resulted in a gradual decrease of total activity, but the ratio of the two enzymes remained essentially constant. No indication was obtained for increased amounts of lysophospholipase I at the expense of lysophospholipase II.

Also SDS-polyacrylamide disc electrophoresis under strongly reducing condi-

TABLE 4. *Effect of ageing of beef liver homogenate at room temperature on the ratio of lysophospholipase I and II*

Time (hr)	Total lysophospholipase (mU)	Lysophospholipase I (mU)	Lysophospholipase II (mU)	Ratio I/II
0	100	11.6	45.2	0.26
1	97	11.0	44.8	0.25
2	92	11.0	42.3	0.26
5	83	11.6	38.0	0.31
22	72	10.2	49.3	0.21

tions gave one protein band for each of the enzymes (Fig. 6). No evidence was obtained for dissociation of the higher molecular weight lysophospholipase II into smaller fragments.

In summary we can conclude that beef liver contains two separate lysophospholipases with different subcellular distribution that do not show a simple mono-

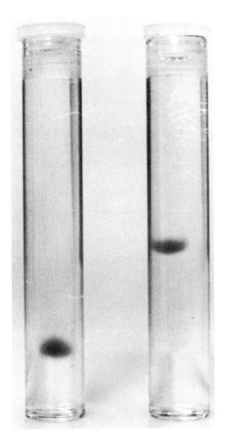

FIG. 6. SDS-disk gel electrophoresis of purified beef liver lysophospholipases. **(Left)** 30 μg of lysophospholipase I. **(Right)** 20 μg of lysophospholipase II.

mer to dimer or trimer relationship. All lysophospholipases isolated so far show a low degree of specificity. Although this may be unfortunate from the point of view of the enzymologist, it does not imply of course a valuation of the importance of these enzymes for the turnover of membrane phosphoglycerides in living cells.

REFERENCES

1. van den Bosch, H., van Golde, L. M. G., and van Deenen, L. L. M. (1972): *Rev. Physiol.,* 66:13.
2. van den Bosch, H., Aarsman, A. J., de Jong, J. G. N., and van Deenen, L. L. M. (1973): *Biochim. Biophys. Acta,* 296:94.
3. de Jong, J. G. N., van den Bosch, H., Aarsman, A. J., and van Deenen, L. L. M. (1973): *Biochim. Biophys. Acta,* 296:105.
4. Krisch, K. (1970): *Z. Klin. Chem. Klin. Biochem.,* 8:545.
5. Sémériva, M., Chapus, C., Bovier-Lapierre, C., and Desnuelle, P. (1974). *Biochem. Biophys. Res. Commun.,* 58:808.
6. Hartley, B. S., and Kilby, B. A. (1954): *Biochem. J.,* 56:288.
7. Piszkiewicz, D., and Bruice, T. C. (1968): *Biochemistry,* 7:3037.
8. Alfonzo, M., and Apitz-Castro, R. (1971): *FEBS Lett.,* 19:235.
9. Feldman, R. I., and Weiher, H. (1972): *J. Biol. Chem.,* 247:267.
10. Pocker, Y., and Stone, J. T. (1968): *Biochemistry,* 7:3021.

Lipids, Vol. 1, edited by R. Paoletti, G. Porcellati, and G. Jacini. Raven Press, New York © 1976.

Characterization and Subcellular Localization of Lipase and Phospholipase A Activities in Rat Liver

P. M. Vignais, A. Colbeau, F. Cuault, H. Ngo-Tri, and M. Pilarska*

DRF/Biochimie, CEN-G., B.P. 85, 38041, Grenoble-Cédex, France

The long-chain fatty acids are essentially found in two types of lipid molecules. In glycerides they represent a reservoir of energy. In phospholipids, components of all biologic membranes, the acyl chains give the membrane its properties of hydrophobicity and fluidity. Lipolytic enzymes such as lipases, able to release the fatty acids stored in glycerides, play a role in the energetic metabolism of the animal cells, in particular in specialized energy-storing cells, the adipocytes, and in cells involved in energy interconversion such as the hepatocytes. Phospholipases A, by altering the lipid components of membranes, may interfere in many ways with the physiology of the cell. Their significance in this respect still remains to be elucidated.

Our approach to a better understanding of the role and mode of action of lipases and phospholipases A was to localize the occurrence of those enzymes in the different compartments of the liver cell, to determine their properties, and then to evaluate their resemblance and possible relationship.

BIOCHEMICAL CHARACTERIZATION OF THE SUBCELLULAR FRACTIONS FROM THE LIVER

The study of the localization of lipases and phospholipases A in the cellular compartments requires first the resolution and identification of subcellular fractions. A critical appraisal of the efficiency of some methods used to fractionate endocellular membranes from rat liver was based on the use of marker enzymes (1). Figure 1 summarizes the methods that we used to isolate from rat liver homogenates different types of organelles. Lysosomes were isolated as Triton WR-1339 filled lysosomes (2) and are therefore referred to as tritosomes. Plasma membranes were isolated by the method of Neville (3). The subcellular fractions

---

* *Present address: Nencki Institute, 3 Pasteur Street, Warsaw, Poland.*

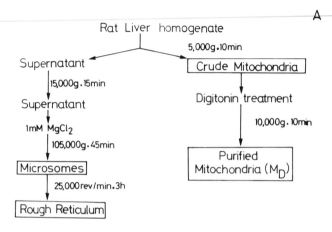

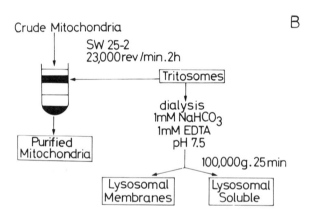

FIG. 1. Flow sheet for purification of mitochondria and isolation of subcellular membranes from normal rat liver **(A)** and from the liver of rats previously injected with Triton WR-1339 **(B)** (cf. ref. 7 for details).

are characterized by the activity of marker enzymes (Table 1), by their lipid composition, and in favorable cases by electron microscopy (1).

As far as the lipid composition is concerned, cardiolipin is considered typical of the inner mitochondrial membrane and sphingomyelin of the plasma membrane. Plasma membrane is particularly rich in cholesterol. Relatively high amounts of sphingomyelin and cholesterol are also found in tritosomes (1). This is probably related to the origin of tritosomes (or secondary lysosomes), which may be formed by pinocytosis of the plasma membrane (4, 5).

A common origin for the plasma membrane and the tritosomal membrane may also explain the presence of 5'-nucleotidase activity in both membranes. However,

TABLE 1. *Identification of membrane fractions with reference to marker enzymes*

Marker enzyme	Mitochondria			Tritosomes		Micro-somes	Plasma Membrane
	Total	Inner Membrane	Outer Membrane	Total	Membrane		
Cytochrome oxidase	2050	4500	600	ND	—	39	80
Monoamine oxidase	6	3	38	ND	—	ND	1
Acid phosphatase	23	ND	170	1040	1930	50	35
NADPH cytochrome c reductase	4	3	30	4	7	94	14
5'-nucleotidase							
pH 5.5	8	—	—	715	1410	—	ND
pH 7.5	7	10	140	560	1310	58	710
pH 9.0	4	—	—	100	440	—	1120

Activities in nmoles·min^{-1}·mg^{-1}. ND = not detected.
From Nachbaur et al., ref. 7.

the optimal activity in the plasma membrane is at pH 9, whereas it is in the acidic range (pH 5) in the tritosomal membrane (6, 7).

CHARACTERIZATION AND DISTRIBUTION OF PHOSPHOLIPASE A ACTIVITIES IN DIFFERENT SUBCELLULAR FRACTIONS

Soluble phospholipase A activities (pH_{opt} 4.5) were found only in tritosomes. These lysosomal acid phospholipases show A_1 and A_2 specificity. Their activity is severalfold higher than that of the alkaline ones, which are membrane-bound, and does not require Ca^{2+}.

Membrane-bound phospholipase A activities can be detected in mitochondria (predominantly A_2), microsomes (predominantly A_1), and plasma membranes with a pH_{opt} in the alkaline range (pH 8.5 to 9.0) (7). In plasma membrane the specificity of attack depends upon the substrate used, predominantly A_1 with phosphatidylethanolamine (PE) and A_2 with phosphatidylglycerol (PG) (7–11). The mitochondrial phospholipase A_2 attacks preferentially endogenous PE to release arachidonic acid (7).

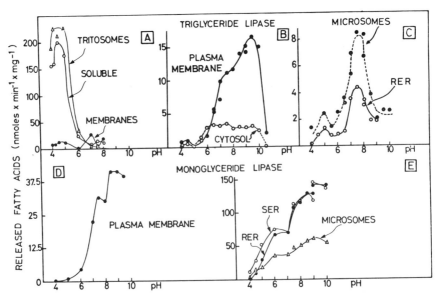

FIG. 2. pH-activity curves for TG-lipase and MG-lipase in different rat liver subcellular fractions. 30 mM triolein or monoolein emulsified in 5% arabic gum. Final volume, 1.5 ml; 30-min incubation at 37°C. **(A)** TG-lipase in tritosomal fractions. Tritosomes, 0.16 mg protein; soluble, 0.19 mg protein; membranes, 0.13 mg protein. **(B)** TG-lipase in plasma membranes, 0.7 mg protein; in cytosol (100,000 g supernatant), 1.4 mg protein. **(C)** TG-lipase in microsomes, 0.7 mg protein; in rough endoplasmic reticulum (RER), 1 mg protein. **(D)** MG-lipase in plasma membrane, 0.19 mg protein **(E)** MG-lipase in microsomal fractions. Microsomes, 0.62 mg protein. Smooth endoplasmic reticulum (SER) and RER, 0.30 mg protein; 20 mM monoolein was used. (From Colbeau et al., ref. 18.)

CHARACTERIZATION AND DISTRIBUTION OF LIPASE ACTIVITIES

The triglyceride-lipase (TG-lipase) activities detected in subfractions of the rat liver cell have been reported as having three different pH_{opt}. As shown in Fig. 2 the acid TG-lipase (pH_{opt} 4.5) is of lysosomal origin and it is found soluble in tritosomes. Alkaline activities are found in microsomal and plasma membrane preparations. TG-lipase is more active in plasma membrane (pH_{opt} 9–9.5) than in microsomes (pH_{opt} 8). The soluble lysosomal TG-lipase has been shown by Teng and Kaplan (12) to require acidic phospholipids for activity. Binding of acidic phospholipids to lysosomal enzymes may contribute to their extraction from the membrane and their acidic specificity.

TG-lipase activity was found in all the fractions containing phospholipase A_1 activity, namely the microsomal, lysosomal, and plasma membrane fractions. On the other hand, mitochondrial preparations from which microsomal, lysosomal, and plasma membrane contaminants had been almost completely removed showed almost no TG-lipase activity (Fig. 3). In agreement with this observation, Assmann et al. (13) detected no lipase activity in highly purified mitochondrial preparations, and Guder et al. (14) and Schousboe et al. (15) concluded that the TG-lipase activity of their mitochondrial preparations was due to the presence of contaminants (lysosomes and microsomes, respectively).

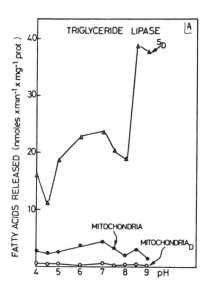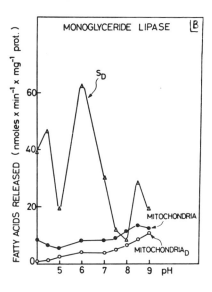

FIG. 3. pH-dependence of TG-lipase activity **(A)** and MG-lipase activity **(B)** before and after purification of a mitochondrial preparation. A rat liver mitochondria suspension was left in contact with digitonin (25 µg digitonin per milligram protein) for 10 min at 0°C and then spun down. The mitochondrial pellet (mitochondria$_D$) and the supernatant fluid were then tested for lipase activities. 10-mM triolein or monoolein emulsified in 5% arabic gum were used. Incubation at 37°C for 60 min **(A)** or 30 min **(B)**. (From Colbeau et al., ref. 18.)

A very active monoglyceride-lipase (MG-lipase) has been detected in smooth and rough endoplasmic reticulum membranes (pH_{opt} 9). MG-lipase activity with a pH_{opt} of 9 is also present in plasma membrane (Fig. 2).

ARE THE LIPASE AND PHOSPHOLIPASE A₁ ACTIVITIES PRESENT IN ENDOPLASMIC RETICULUM AND PLASMA MEMBRANES DUE TO THE SAME ENZYME?

The following investigations were initiated by the reports (16, 17) that purified lipase preparations from porcine pancreas and from the mold *Rhizopus arrhizus* may have a weak phospholipase A_1 activity, i.e., hydrolyze the fatty acid ester bond at the 1-position of the PGs. In an attempt to determine whether the lipase and phospholipase A_1 activities present in microsomal and plasma membranes are due to the same enzyme, we have examined the effect of heat denaturation or of detergent and salt extraction on these activities (18). In microsomes and plasma membranes, the TG-lipase and phospholipase A activities are altered differently by heat denaturation, delipidation, or KCl extraction. TG-lipase and phospholipase A are more easily extracted by salt from the membrane than is MG-lipase.

These preliminary data suggest, although they do not prove, that the TG-lipase, the MG-lipase, and the phospholipase A_1 are different protein entities.

ISOLATION OF TG-LIPASE AND MG-LIPASE BY GEL FILTRATION

Evidence for a single protein having multiple enzyme properties requires the isolation and purification to homogeneity of the lipase and phospholipase A_1 activities. Working on the same problem, Van den Bosch et al. (19) have recently shown that the highly purified phospholipase A_1 of beef pancreas can also exhibit lysophospholipase activity and under certain conditions phospholipase B activity. They also found that phospholipase A_1 was different from pancreatic lipase. Likewise, in their purified preparation of rat liver lysosomal TG-lipase, Teng and Kaplan (12) did not detect any phospholipase A activity. On the other hand, Waite and Sisson (20) concluded that the phospholipase A_1 solubilized by heparin from the plasma membrane of rat liver also exhibits lipase activity since it can catalyze the hydrolysis of MGs and DGs. Furthermore this enzyme was found to exhibit transacylase activity since it was able to transfer the acyl group from the 1-position to an acceptor MG to form a DG.

We have chosen to compare the lipolytic activities present in a microsomal fraction, namely the phospholipase A, TG- and MG-lipase activities, and the transacylase activity using 1-monoglyceride as acceptor. By salt extraction TG-lipase and phospholipase A are extracted preferentially to MG-lipase from the microsomes (18). However, since MG-lipase is very active, enough MG-lipase is found in the soluble extract to allow the monitoring of the enzyme in the gel filtration fractions.

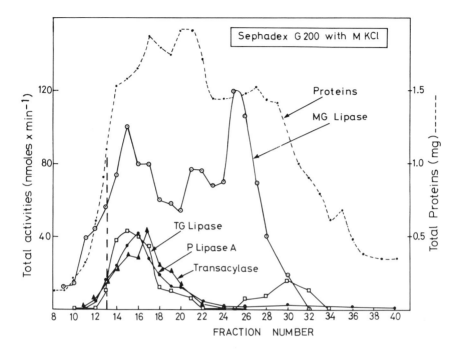

FIG. 4. Gel-filtration pattern of a KCl extract of microsomes. Rat liver microsomes were extracted for 10 min at 0°C with 0.5 M KCl and centrifuged. The supernatant fraction was dialyzed overnight against water, vacuum-concentrated, made 1 M with KCl and filtrated on a column of Sephadex G-200 which had been equilibrated with 1 M KCl and 50-mM Tris-HCl, pH 8.5. Fractions of 1 ml were collected from which 0.2-ml aliquots were used for TG-lipase assay and about 0.1 ml for each other assay. Dashed line: elution volume of blue dextran.

Phospholipase assay **(P Lipase A, ●):** 0.1-mg protein was incubated in 1 ml of 0.01 M Tris-HCl buffer, pH 9.0, 0.12-M KCl, 0.002-M CaCl₂, and 400-μM (³²P)PE for 10 min at 37°C. The amount of (³²P)lyso-PE formed is determined after thin-layer chromatography (TLC) separation of PE and lyso-PE.

Transacylase assay (▲): Approximately 0.1-mg protein was incubated in 1 ml of 0.01-M Tris-HCl buffer, pH 9.0, 0.12-M KCl, 0.002-M CaCl₂, 100-μM 1-[¹⁴C]monopalmitine (spec. radioactivity 100 dpm × nmole⁻¹). Incubation time, 10 min at 37°C. The amount of [¹⁴C]diglyceride formed was determined after TLC separation of the glycerides.

The lipase assays were carried out in a total volume of 1.5 ml containing 0.05-M Tris-HCl buffer, pH 8. Incubation at 37°C for 10 min. The released fatty acids were measured by the method of Duncombe (21) as modified by Mahadevan et al. (22). TG Lipase (□): 0.2 ml of each fraction and 30-mM triolein emulsified in 5% arabic gum were used. MG-lipase (○): 0.1-ml of each fraction and 20-mM monoolein emulsified in 5% arabic gum.

In the first series of experiments, microsomes were extracted by 0.5 M KCl, and after dialysis and vacuum-concentration the extract was made 1 M with KCl and filtrated on a Sephadex G-200 column equilibrated with 1 M KCl and 50 mM Tris-HCl, pH 8.5. The elution pattern (Fig. 4) indicates that the four tested activities are found in a peak very soon after the exclusion peak; however, whereas the phospholipase and transacylase activities are found together only in that early peak, the MG-lipase activity (peaking in the 25th fraction) and the TG-lipase activity (peaking in the 30th fraction) could be further chromatographed and separated.

Assuming that the first peak at which the lipolytic activities were all found could be due to a protein aggregate, in the next series of experiments Triton X-100 was included in the eluting solvent (Fig. 5). However, the amount of Triton X-100 (0.05 to 1%) was maintained low enough to avoid interference with the determination of the TG-lipase and phospholipase A activities (18) and of transacylase activity (20). In this case again, the MG-lipase and TG-lipase activities could be separated from the rapidly eluted peak, although they were not as clearly separated from one another as in the previous experiment (Fig. 4); the phospholipase A and the transacylase activities remained both together in the first peak. Subsequently fractions 12 to 20 of the experiment described in Fig. 5 were pooled, concentrated under vacuum, and filtrated again on a Sephadex G-200 column that had been equilibrated with 50 mM Tris-HCl, pH 8.5, 0.1 M KCl, and 1% Triton X-100. Elution with the same mixture gave rise to a main peak where all the tested activities were found followed by a second peak (peaking at the 40th fraction) where the specific activity of MG-lipase was the highest (not shown).

These preliminary results indicate that it is possible to isolate fractions having only MG-lipase or TG-lipase activity and where no phospholipase A activity is detected. On the other hand, in the rapidly extruded peak, phospholipase A activity and transacylase activity are closely associated. As mentioned above gel filtration may not have allowed a true separation of the components but on the contrary may have brought about their aggregation. However, a deacylating enzyme can reasonably be expected to carry a transacylation reaction if an acceptor molecule more hydrophobic than water such as MG can easily approach the active center presumably imbedded in a hydrophobic region. If the first peak coming out very close to the breakthrough peak is made of an aggregate one should expect to have a good transacylase activity maintained. Membrane proteins show a strong affinity for lipids, and it will be interesting to determine if specific lipids are associated with the proteins present in the first peak, and if it is thanks to the occurrence of transacylase activity that those lipids are apparently not hydrolyzed. Lipases and phospholipases A offer interesting examples of lipid-protein interactions to study. The careful use of detergents as described by Helenius and Simons (23) or of more sophisticated extraction procedures such as the phase-partition method set up by Albertsson (24) to purify membrane-bound phospholipase A_1 from *Escherichia coli* should lead to a better understanding of the lipid-protein interactions involved.

Waite and Sisson (20) using heparin extracts of the rat liver plasma membrane have reported that phospholipase A_1 and transacylase activities co-eluted together from a column of Sephadex G-200. Our results extend those of Waite and Sisson (20) to enzymes originating from the endoplasmic reticulum. The data may lead to a more general understanding of the mode of action of the lipolytic enzymes. On the other hand, they warn us to avoid to attribute to enzymes a too specific physiologic role restricted to the membrane of origin. Finally it may

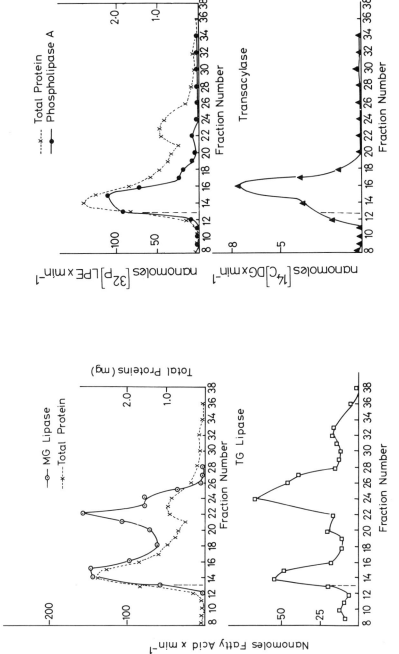

FIG. 5. Gel filtration pattern of a KCl extract of microsomes. Same extraction conditions as in Fig. 4. After dialysis and concentration the extract was made to 50 mM Tris, HCl, pH 8.5, 0.1 M KCl and 0.05% Triton X-100 and filtrated on a column of Sephadex G-200 which had been equilibrated with the same Tris-KCl-Triton X-100 mixture. Dashed line: elution volume of blue dextran.

not be so surprising to find similar enzymes in both the endoplasmic reticulum and plasma membranes if, as suggested by Morré et al. (25), plasma membrane is derived from endoplasmic reticulum by a dynamic flow and differentiation.

ACKNOWLEDGMENTS

This work was supported by research grants from the CNRS (ERA no. 36) and the INSERM.

REFERENCES

1. Colbeau, A., Nachbaur, J., and Vignais, P. M. (1971): *Biochim. Biophys. Acta,* 249:462.
2. Wattiaux, R., Wibo, M., and Baudhuin, P. (1963): In: *Lysosomes,* edited by A. V. S. Renck, and M. P. Cameron, p. 176. Ciba Foundation Symp., Churchill, London.
3. Neville, D. M., Jr. (1968): *Biochim. Biophys. Acta,* 154:540.
4. de Duve, C., and Wattiaux, R. (1966): *Ann. Rev. Physiol.,* 28:435.
5. Henning, R., and Heidrich, H. G. (1974): *Biochim. Biophys. Acta,* 345:326.
6. Kaulen, H. D., Henning, R., and Stoffel, W. (1970): *Hoppe Seylers Z. Physiol. Chem.,* 351:1555.
7. Nachbaur, J., Colbeau, A., and Vignais, P. M. (1972): *Biochim. Biophys. Acta,* 274:426.
8. Torquebiau-Colard, O., Paysant, M., Wald, R., and Polonovski, J. (1970): *Bull. Soc. Chim. Biol.,* 52:1061.
9. Victoria, E. J., van Golde, L. M. G., Hostetler, K. Y., Scherphof, G. L., and van Deenen, L. L. M. (1971): *Biochim. Biophys. Acta,* 239:443.
10. Newkirk, J. D., and Waite, M. (1971): *Biochim. Biophys. Acta,* 225:224.
11. Newkirk, J. D., and Waite, M. (1973): *Biochim. Biophys. Acta,* 298:562.
12. Teng, M., and Kaplan, A. (1974): *J. Biol. Chem.,* 249:1064.
13. Assmann, G., Krauss, R. M., Fredrickson, D. S., and Levy, R. I. (1973): *J. Biol. Chem.,* 248:1992.
14. Guder, W., Weiss, L., and Wieland, O. (1969): *Biochim. Biophys. Acta,* 187:173.
15. Schousboe, I., Bartels, P. D., and Jensen, P. K. (1973): *FEBS Lett.,* 35:279.
16. de Haas, G. H., Sarda, L., and Roger, J. (1965): *Biochim. Biophys. Acta,* 106:638.
17. Slotboom, A. J., de Haas, G. H., Bonsen, P. P. M., Burbach-Westerhuis, G. J., and van Deenen, L. L. M. (1970): *Chem. Phys. Lipids,* 4:15.
18. Colbeau, A., Cuault, F., and Vignais, P. M. (1974): *Biochimie,* 56:275.
19. van den Bosch, H., Aarsman, A. J., and van Deenen, L. L. M. (1974): *Biochim. Biophys. Acta,* 348:197.
20. Waite, M., and Sisson, P. (1973): *J. Biol. Chem.,* 248:7985.
21. Duncombe, W. G. (1963): *Biochem. J.,* 88:7.
22. Mahadevan, S., Dillard, C. J., and Tappel, A. L. (1969): *Anal. Biochem.,* 27:387.
23. Helenius, A., and Simons, K. (1972): *J. Biol. Chem.,* 247:3656.
24. Albertsson, P. Å. (1973): *Biochemistry,* 12:2525.
25. Morré, D. J., Franke, W. W., Deumling, B., Nyquist, S. E., and Ovtracht, L. (1971): In: *Biomembranes,* Vol. 2, edited by L. A. Manson, p. 95. Plenum Press, New York.

Lipids, Vol. 1, edited by R. Paoletti, G. Porcellati,
and G. Jacini. Raven Press, New York © 1976.

Mode of Action of the Plasmalemma Phospholipase from Rat Liver

Moseley Waite and Patricia Sisson

*Agricultural Research Council, Institute of Animal Physiology, Babraham, Cambridge CB 2 4AT
England*

Over the past several years, our group has been involved in the study of phospholipases of the liver, their subcellular localization, their purification, and their function (1–9). In this volume, Dr. Vignais presented work summarizing the subcellular localization of these enzymes, and I shall therefore concentrate in this chapter on the unusual characteristics of one of these enzymes and propose a role for this enzyme in the uptake of lipid by the liver. This enzyme, which is located in the plasmalemma, is specific for position 1 of phospholipid and therefore was originally defined as a phospholipase A_1.

The first part of this project centered on obtaining purified plasmalemma preparations that were characterized by their marker enzyme content. Using either rate-zonal isopycnic centrifugation (4) or the procedure of Nelville (10) we obtained preparations that had low contamination with other organelles. We then compared the specific activity of the phospholipases A_1 of the plasmalemma and of the similar phospholipase A_1 in microsomes with that of marker enzymes (5′-nucleotidase and NADPH cytochrome c oxidase) and found that each organelle did contain the phospholipase A_1. Thus far we have been unable to distinguish between the two enzymes on a basis other than their subcellular localization and the tightness by which they are associated with the membranes. This enzyme, when associated with the membranes, was optimally active at pH 9.0 to 9.5, required Ca^{2+}, and of the phospholipids tested preferred phosphatidylethanolamine (PE). In these studies we also found that both organelles contained a phospholipase A_2, which preferentially hydrolyzed phosphatidylglycerol (6), in agreement with the findings of Victoria et al. (11).

Shortly after these early observations, Zieve and Zieve (12) demonstrated that the phospholipase A_1 of postheparin serum was absent in rat whose livers had been removed. This was not the case with the lipoprotein lipase. They concluded on this basis that the serum phospholipase A_1 was released into the blood from the liver upon injection with heparin. Furthermore, they suggested that this could be the same enzyme we had described in the plasmalemma of liver. Based on these observations we undertook the project of investigating the influence of heparin on the isolated plasmalemma and the role of the nature of the reaction catalyzed by the enzyme.

We first studied the effect of heparin on the phospholipase and found that low concentrations of heparin (10 to 50 μg/ml) stimulated the hydrolysis of PE two- to threefold under a variety of conditions (7). This stimulation was blocked by protamine, which binds heparin. Furthermore, other sulfated mucopolysaccharides such as the chondroitin sulfates were poor substitutes for heparin. If the plasmalemma preparations were incubated with heparin and the membranes sedimented by centrifugation, most of the phospholipase A_1 was solubilized (Fig. 1). The low amount of enzyme solubilized without the addition of heparin could have originated from the small amount of contaminating microsomes, since in other experiments it was demonstrated that the enzyme was easily washed from microsomes. We found that the same amount of heparin (10 μg/mg plasmalemma protein) caused maximal solubilization and stimulation of the enzyme and was equivalent to the amount of (^{35}S) heparin that could be bound by the plasmalemma. These results suggest that there is a specificity in the heparin-dependent release of the phospholipase A_1 from the plasmalemma.

Although Zieve et al. (13) had concluded that the serum phospholipase A_1 and lipase were separate enzymes, we felt that our plasmalemma system was better

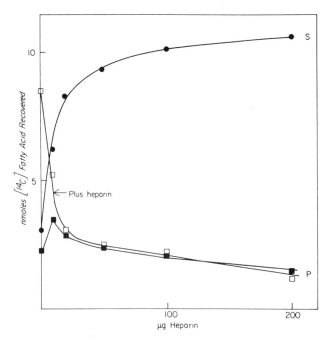

FIG. 1. Displacement of the phospholipase A_1 from the plasma membrane by heparin. Five milligrams of plasma membrane protein were incubated with the indicated amounts of heparin as described in text. The assay mixtures for phospholipase A_1 activity contained 0.10 ml of the soluble (●) or resuspended particulate fractions. The particulate fraction was assayed either without the addition of heparin to the assay mixture (■) or in the presence of 50 μg of heparin (□).

defined than serum for the examination of substrate specificity of the phospholipase A_1. To isolate the enzyme, we first washed the plasmalemma with isotonic saline, which removed about 35% of the protein. Following this we eluted the phospholipase A_1 with a rather high specific activity. With such a preparation we found that the enzyme could hydrolyze sn-glycerol-3-phosphorylethanolamine (monoacyl-GPE), monoacylglycerol, and diacylglycerol in addition to diacyl-GPE (Table 1). Indeed, it appeared as if monoacylglycerol was the preferred substrate. There are two other interesting points to be observed in Table 1; first, when monoacylglycerol was present, diacylglycerol was produced, and, second, Triton X-100 reduced the production of diacylglycerol while either having no effect (with monoacylglycerol as substrate) or stimulating (with diacylglycerol as substrate) hydrolysis. We found that the enzyme could not incorporate free fatty acid under conditions that lead to transacylation (Table 2). The addition of the cofactors associated with reacylating activity, adenosine triphosphate (ATP), coenzyme A (CoA), and Mg^{2+}, had no influence. We therefore concluded that the enzyme removes an acyl group from a donor (reactions 1, 2, and 5 in Fig. 2) and can directly attach it to an acyl acceptor, monoacylglycerol (reactions 3 and 4). The other acyl acceptor shown here is H_2O, the product of the fatty acid. This process, which appears to have an acyl-enzyme intermediate, is termed transacylation. Whether the enzyme will use the hydroxyl of monoacylglycerol or that of H_2O depends upon the availability of the acceptor hydroxyl to the

TABLE 1. *Composition of the three Lipoprotein fractions obtained by molecular sieving[a]*

Substrate	Triton	Added lipid	Product (nmoles)		
			MG	DG	FA
1. 1- 9,10-[³H₂]PE	−			0	7.5
	−	MG		2.7	3.0
	−	DG		1.2	3.0
	+			0	4.2
	+	MG		0	2.6
	+	DG		0	2.4
2. 1- 9,10-[³H₂]MAGPE	−			0	8.8
	−	PE		0	1.9
	−	DG		1.9	3.7
	−	MG		1.8	0.4
3. 1- 1-[¹⁴C]MG	−			19.4	9.9
	−	PE		15.9	9.8
	+			3.2	7.7
4. 2- 1-[¹⁴C]DG	−		6.4		3.0
	+		16.2		0.9

The values of the lipoprotein run without addition are the averages from four determinations; those with the added labeled lipids are from 10 determinations. MG, monoglyceride; DG, diglyceride; FA, fatty acid; PE, phosphatidylthanolamine; MAGPE, monoacylglycerophosphorylethanolamine.

TABLE 2.

Substrate	3H				14C			
	MG	1,2-DG	1,3-DG	FA	MG	1,2-DG	1,3-DG	FA
1. 9,10-[³H₂]oleic acid + 1- 1-[¹⁴C]MG	0	0.1	0	0		7.1	12.3	14.6
2. 9,10-[³H₂]oleic acid + 1- 1-[¹⁴C]MG + PE	0.03		0.02			5.9	9.0	13.6
3. 1-[¹⁴C]linoleic acid +1- 9,10-[³H₂]PE + MG		1.4	1.4	2.8	0	0	0.03	

Product (nmoles)

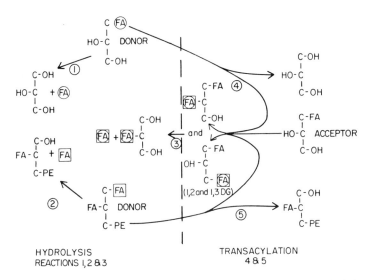

FIG. 2. The fate of the 1-acyl group from monoglyceride or from PE in the hydrolytic reactions 1 to 3 **(left of the dotted line)** and the transacylation reactions 4 and 5 **(right of the dotted line).** In this scheme **FA** designates fatty acid, **DG,** diglyceride, and **-PE** denotes the phosphorylethanolamine moiety of phosphatidylethanolamine.

enzyme. Triton, which forms hydrated micelles (14), brings H_2O into the substrate liposome and causes a reduction in the transacylation but not in the hydrolysis. If, however, we used concentrations of Triton higher than 0.05% all activities were inhibited, which suggests that an organization of the lipid such as that found in the liposome is required for enzymatic activity.

The interesting characteristic of this enzyme, which allowed it to utilize acyl acceptors other than water, prompted us to investigate the requirements of the enzyme for both the acyl donor and the acyl acceptor molecules. First we examined the potential of the enzyme to utilize the acyl group of position 2 of monoacylglycerol, the preferred substrate. This problem was complicated by the tendency of the acyl group at position 2 to migrate to position 1 of the primary alcohol moieties. We therefore compared the activity of the phospholipase A_1 with that of the pancreatic lipase, an enzyme known to be specific for positions 1 and 3 of acylglycerols.

As seen in Fig. 3 approximately 50% of the [^{3}H]monoacylglycerol (prepared as the 2-acyl isomer) was hydrolyzed by the lipase which demonstrates that 50% of the molecules had undergone acyl migration. The complete hydrolysis of the 1-[^{14}C]monoacylglycerol demonstrates that sufficient lipase was used to totally break down the monoacylglycerol used. The results in the bottom half of the figure show that the phospholipase A_1, similar to the lipase, is not capable of utilizing the acyl group at position 2. In similar experiments we demonstrated that the phospholipase cannot utilize the acyl group at position 2 of monoacyl-GPE. Thus far we have been unable to demonstrate any activity on triacyl-

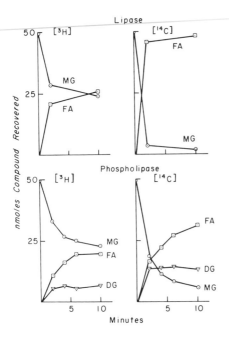

FIG. 3. Utilization of 2-[³H]arachidonoyl-glycerol **(left)** and 1-[¹⁴C]oleoylglycerol **(right)** by pancreatic lipase **(upper)** and phospholipase A₁ **(lower).** Each substrate (50-μM which was added last as an ultra-sonic suspension prepared in 0.05% Triton X-100) was incubated for the indicated times in 1.0 ml of 100-mM Tris-HCl, pH 7.0, with 5-mM CaCl₂ and either 200 μg of pancreatic lipase or 176 μg of the solubilized protein from the plasma membrane. (□) Fatty acid, (○) monoacylglycerol, (△) diacylglycerols.

glycerol, which suggests that this enzyme has no lipase activity. A summary of our studies on substrate specificity are presented in Table 3.

In contrast to the requirements for the acyl donor site, there is little restriction upon the acyl acceptor molecule. The acyl group can go into position 2 of monoacylglycerol, 1-acyl-1,2 propanediol, acylglycol, or 1-alkylglycerol. It does appear, however, that there is a preference for the primary hydroxyl. The hydroxyl group need not be adjacent to the acyl or alkyl chain, as shown by the acylation of 1-acyl, 1,3 propanediol and the production of 1-alkyl-3-acylglycerol from 1-alkylglycerol. We found that the enzyme could also acylate cetyl alcohol

TABLE 3. *Substrates for phospholipase A₁*

Acyl donors	Acyl acceptors
1-acylglycerol	1- and 2-acylglycerol
1,2- and 1,3-diacylglycerol	1-acylpropanediols
1-acyl-GPE	acylglycol
diacyl-GPE	1- and 2-alkylglycerols
	H₂O
	cetyl alcohol
Not acyl donors	Not acyl acceptors
2-acylglycerol	diacylglycerol
2-acyl-GPE	1-acyl- or 2-acyl-GPE
triacylglycerol	cholesterol

at a rather rapid rate but could not utilize the hydroxyl of cholesterol. Furthermore the enzyme could not utilize diglyceride nor either of the isomeres of monoacyl-GPE. These results were obtained using both monoacylglycerol and diacyl-GPE as donors. On this basis we conclude that the only restriction on the utilization of a compound as acyl acceptor is size; those compounds that are bulky cannot fit into the acyl acceptor site.

There are two possibilities to account for the differences in the recognition of acyl acceptors and donors by the enzyme. First, there could be two distinct sites on the enzyme, one for the donor and one for the acceptor molecules. Presumably these two sites would have in common the active site moiety that becomes acylated. Second, it is possible that there is a single site on the enzyme that changes in character upon acylation of the enzyme. Obviously the presence of the acyl group in the active site will restrict the entry and utilization of the acceptor molecule notably. In addition there could be a conformational change in the enzyme upon acylation. At present we believe either that there must be two sites on the enzyme or that a conformational change must take place within a single active site. Otherwise we cannot account for the differences found in the positional requirements for the acyl donor and the size limitation for the acyl acceptors. This is best demonstrated by the observation that a compound that is a product of the acyl donation, monoacyl GPE, cannot be used as an acyl acceptor.

With this knowledge concerning the substrate specificity of the enzyme, we then turned our studies to the physiologic role of the enzyme. Stein and Stein (15) earlier suggested that a phospholipase could be involved in the uptake of phospholipid by the liver. Furthermore, it was shown by Redgrave (16) that the remnant lipoprotein derived from chylomicra and very low-density lipoproteins could be cleared by the liver. The remnant lipoproteins, which had much of the triglyceride hydrolyzed by the extrahepatic lipoprotein lipase, have a high content of phospholipid, cholesterol-ester, and apoprotein. This again suggests that a phospholipase could be one of the enzymes responsible for uptake of lipoproteins by the liver. Based on the various activities we have previously described for the phospholipase, we formulated as a working model the scheme outlined in Fig. 4.

In this scheme the phospholipase A_1 takes monoacylglycerol produced by the lipoprotein lipase as a central carrier molecule that accepts acyl groups from phospholipid, diglyceride (DG), or monoglyceride (MG). Because this enzyme normally would function within the membrane, it is possible that the product DG could become associated with the membrane and even pass to the cell interior, thereby serving as a lipid transporter. The diacylglycerol then would be converted directly to phospholipid or triacylglycerol. The monoacylphosphoglyceride would be taken into the membrane and reacylated in a reaction that employs acyl CoA (17). Such a mechanism would have obvious energetic advantages to the cell in the uptake of both phospholipid and the remainder of the neutral glycerides, monoacylglycerol. To actually demonstrate such a process as physiologically meaningful will of course be quite difficult and at the offset, open to

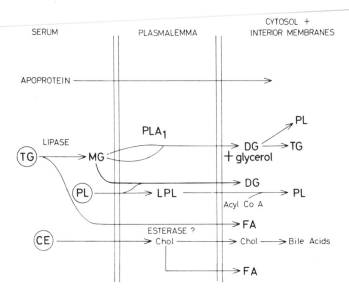

FIG. 4. Possible clearance of remnant lipoproteins by liver. **TG,** triacylglycerol; **DG,** diacylglycerol; **MG,** monoacylglycerol; **PL,** phospholipid; **CE,** cholesterol ester; **Chol,** cholesterol; **FA,** fatty acid; **LPL,** monoacyl phosphoglyceride; **PLA,** phospholipase A_1.

many questions. We believe, however, that a systematic approach toward this problem will be quite important, no matter how much modification our original ideas must undergo. Initially we have asked two questions; first, what reactions can the phospholipase A_1 catalyze when the lipid substrate is a part of serum lipoproteins, and second is there a specificity in uptake of the reaction products by the plasmalemma?

We found that the addition of a mixture of serum lipoproteins decreases the apparent enzymatic activity on liposomes of a mixture of radioactive monoacylglycerol and diacyl-GPE. This could have been the result of inhibition of activity by the lipoproteins, which would suggest that they are not substrates, or this could have been the result of isotopic dilution, which would suggest that the enzyme was active on both forms of the lipid. It was necessary therefore to devise a system in which we could label the lipid of the lipoprotein and remove from the liposomes the nonincorporated lipid of the liposomes. Using the gel filtration technique of Rudel et al. (18), we developed a system that appears to be a good model of the lipoproteins isolated from serum. Basically we incubated a mixture of lipoproteins (concentrated by ultracentrifugation) with liposomes containing labeled [^{14}C]monoacylglycerol and [^{3}H]diacyl-GPE of high specific activity. This mixture was then passed through a column of Agarose.

Figure 5 shows the elution pattern of the radioactive lipid; the absorbancy at 280 nm was similar. Plotted above each peak is the isotopic ratio of the peak area. Only with peak A is there an appreciable alteration in the isotopic ratio. This was due to the presence of liposomes of [^{3}H]diacyl-GPE that had not been

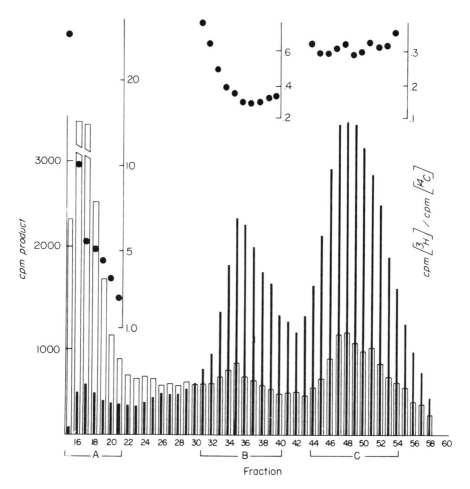

FIG. 5. The separation of labeled **(A)** very low-density, **(B)** low-density, and **(C)** high-density lipoproteins on a column of Biogel A15. Two ml of the lipoprotein concentrate was incubated with 2.0 ml of a sonicate that contained 1.4 μmoles of 1-[^{14}C]oleoylglycerol (1.7 X 10^6 cpm per μmole) and 0.7 μmoles of 1-[^{3}H]diacyl-GPE (9.7 X 10^6 cpm per μmole) as described in the text. The mixture was filtered through the column and collected in 2.5-ml aliquots. One-fifth ml of each was counted in a scintillation counter and the absorbancy at 280 nm determined. The remainder of each tube was then pooled as indicated in the figure and centrifuged. The amount of 1-[^{14}C]oleoylglycerol recovered was 1.5 X 10^6 cpm (64%) and that of 1-[^{3}H]diacyl-GPE 8.1 X 10^5 cpm (12%).

absorbed by the lipoprotein. Ultracentrifugation removed these and gave a lipoprotein fraction with a much lower isotopic ratio. Quantitation of the lipoprotein fraction showed that little change had occurred in the chemical composition of the lipoproteins. We therefore believe that these three fractions, A, B, and C, are good models for the study of the very low-density, low-density, and high-density lipoproteins, respectively. The amount of labeled lipid added amounts to only 2 to 5% of the lipid present in the lipoprotein mixture used.

TABLE 4. *Specific activity of the phospholipase*

Experiment I	1-[³H]diacyl-GPE (nmoles/mg/15 min)			1-[¹⁴C]oleoylglycerol (nmoles/mg/15 min)		
	A	B	C	A	B	C
Plasma membranes	17	5	5	75	60	65
Soluble fraction	16	8	6	44	50	17

Incubation mixtures contained either 25 µg of soluble protein or 50 µg plasma membrane protein and were incubated for 10 min. (A) Very low-density, (B) low-density, and (C) high-density lipoproteins.

The specific activities of the plasmalemma and the soluble fraction obtained from plasmalemma are compared in Table 4. Since it is not possible to vary the concentration of one lipid independent of the others and since the enzyme is active on more than one lipid, we cannot consider these values as true specific activities. Nonetheless we can contrast the activities found using the two sources of enzyme on the three fractions of lipoproteins. Two points of interest are to be noted in these data. First, the plasmalemma utilizes the monoacylglycerol in the three fractions to an equal extent whereas the activity on diacyl-GPE is three times higher with fraction A than with either B or C. Second, the soluble fraction and the plasmalemma have similar activity on diacyl-GPE whereas the soluble fraction has lower activity on monoacylglycerol, especially in fraction C. We conclude that the phospholipase is capable of utilizing the lipid in lipoprotein and that the lipid in the very low-density lipoprotein (A) is preferred. The plasma membranes could also contain a second enzyme that is not released by heparin that is active on monoacylglycerol. This presumed enzyme would be most active on the high-density lipoprotein (C).

The product of these reactions is mainly fatty acid. The low production of diacylglycerol could be the result of the lipoprotein providing an unfavorable physical environment for transacylation, perhaps due to increased hydration of the lipid body (compared with the liposome). This could also be the result of the low amount of monoacylglycerol present, about 10 to 20% of that used in the liposomes. To test the second hypothesis we incubated liposomes of nonlabeled monoacylglycerol with labeled lipoproteins. This caused a two- to threefold increase in the production of diacylglycerol and a concomitant decrease in hydrolysis. It is possible *in situ* that there is a continual generation of monoacylglycerol by the action of the lipoprotein lipase, which would function as the acyl acceptor we visualize functioning in the transport process (Fig. 4).

We investigated the ability of the plasmalemma to bind the reaction products by incubating the membrane or fractions derived by heparin treatment with the lipoproteins, sedimenting the membranes by centrifugation and determining the amount of each compound in both the soluble and particulate fractions.

The data in Table 5 show that only a little of the radioactive lipid from the

TABLE 5. *Uptake of lipids and their reaction products by the plasma membranes*

Substrate	Enzyme fraction	Percent of label in pellet	Product			
			In soluble		In pellet	
			cpm	%	cpm	%
1-[^{14}C]oleoylglycerol	Plasma membranes (no incubation)	11.1	267	22.3	70	47.1
	Plasma membranes	41.1	349	45.3	455	70.1
	Soluble	5.1	452	26.8	24	27.5
	Pellet	22.9	185	16.4	137	39.3
	Soluble + pellet	32.8	401	32.1	654	55.5

The indicated amounts of protein were incubated for 2 min at 37°C with lipoprotein fraction A (2.6 × 10⁴ cpm of [^{14}C]). The reaction mixtures were chilled in ice and centrifuged at 10,000 ×g for 10 min. The pellet was resuspended in 1.0 ml of the original buffer and 0.1 ml of the soluble and pellet fractions were counted in a scintillation counter. The remaining 0.9 ml was extracted and chromatographed to quantitate the products formed. The percent in the pellet was calculated by dividing the cpm in the pellet by the total of the cpm recovered in both fractions. The percentage product in each fraction was determined by dividing the cpm of that compound by the total radioactivity recovered from the thin-layer plate. The cpm of each compound was calculated as the product of the total cpm in the fraction (soluble or particulate) and the percentage of that compound.

very low-density lipoprotein (A) becomes associated with the membrane without incubation at 37°C (11%) or if the membranes were treated with heparin to remove the phospholipase A_1 (22.9). Nearly half of the radioactive lipid became associated with the membranes upon incubation for 2 min at 37°C. If the amount of membrane protein was increased from 0.10 mg to 0.40 mg, about 80% of the radioactive lipid was associated with the membrane. To demonstrate that the phospholipase A_1 is responsible for the uptake of lipid from the lipoprotein, we added the soluble fraction back to the heparin-treated membranes. This caused a 50% increase in the lipid uptake (22 versus 33%). When the radioactive lipid in the resuspend pellets was chromatographed, we found that the membranes contained labeled diacylglycerol and fatty acid mainly (70% with intact membranes and 55% with the soluble plus heparin-treated membranes). The actual counts per minute of product increased considerably if the amount of membrane protein was increased. Similar results were obtained when we studied the uptake of 1-[^{3}H]diacyl-GPE in the very low-density lipoprotein (A) by the membrane. The uptake from lipoprotein fractions B and C was somewhat lower than with fraction A. This is thought to be the result of lower enzymatic activity on fractions B and C.

Based on these observations we conclude that our model for the uptake of lipid by the plasmalemma (Fig. 4) merits further experimentation. There are many problems yet to be answered. One obvious question involves the isomer of the monoacylglycerol that is involved in the transacylation. Another is the necessity to show uptake of monoacylglycerophosphatides by the plasmalemma, perhaps with a concomitant reacylation process.

Another basic question is obvious: why is this enzyme termed a phospholipase? As is often the case, the answer is historical. We now recommend that the enzyme be renamed "monoacylglycerol transacylase." This is based on what has been found to be the preferred substrate, monoacylglycerol, and what is the preferred reaction, transacylation. We believe that as we learn more about the activity of this enzyme *in situ,* more evidence will be obtained for a physiologically significant role for the transacylation physiologically.

By way of a footnote, we have tried several approaches to demonstrate that a single enzyme catalyzes the various reactions described and that this enzyme is the same as the phospholipase A_1 of serum. We found that all activities coeluted from a column of Sephadex G-200, all were inactivated to the same extent by heating, and that all had the same stability upon storage. Furthermore, we were able to demonstrate all of the activities described here using the monoacylglycerol transacylase purified to homogeneity from human postheparin serum.

ACKNOWLEDGMENTS

This work was supported in part by U.S. Public Health Service Grant AM-11799 and General Research Support Grant RR-5404 of the National Institutes of Health. M.W. is U.S. Public Health Service Career Development Awardee 17392.

The human postheparin serum enzyme preparation was kindly donated by Drs. Fruede and Zieve of the Veteran's Administration Hospital, Minneapolis, Minnesota.

REFERENCES

1. Waite, M., and van Deenen, L. L. M. (1967): *Biochim. Biophys. Acta,* 137:498.
2. Waite, M., Scherphof, G. L., Boshouwers, F. M. G., and van Deenen, L. L. M. (1969): *J. Lipid Res.,* 10:411.
3. Waite, M. (1969): *Biochemistry,* 8:2536.
4. Newkirk, J., and Waite, M. (1971): *Biochim. Biophys. Acta,* 225:224.
5. Franson, R., Waite, M., and LaVia, M. (1971): *Biochemistry,* 10:1942.
6. Newkirk, J., and Waite, M. (1973): *Biochim. Biophys. Acta,* 298:562.
7. Waite, M., and Sisson, P. (1973): *J. Biol. Chem.,* 248:7201.
8. Waite, M., and Sisson, P. (1973): *J. Biol. Chem.,* 248:7985.
9. Waite, M., and Sisson, P. (1974): *J. Biol. Chem.,* 249:6401.
10. Nelville, D. M., Jr. (1968): *Biochim. Biophys. Acta,* 154:540.
11. Victoria, E. G., van Golde, L. M. G., Hostetler, K. Y., Scherphof, G. L., and van Deenen, L. L. M. (1971): *Biochim. Biophys. Acta,* 239:443.
12. Zieve, F. J., and Zieve, L. (1972): *Biochem. Biophys. Res. Commun.,* 47:1480.
13. Zieve, F. J., Fruede, K. A., and Zieve, L. (1973): *Fed. Proc.,* 32:561.
14. Kushner, L. M., and Hubbard, W. D. (1954): *J. Phys. Chem.,* 58:1163.
15. Stein, Y., and Stein, O. (1966): *Biochim. Biophys. Acta,* 116:95.
16. Redgrave, T. G. (1970): *J. Clin. Chem.,* 49:465.
17. Lands, W. E. M., and Hart, P. (1965): *Biol. Chem.,* 240:1905.
18. Rudel, L. L., Lee, J. A., Morris, M. D., and Felts, J. M. (1974): *Biochem. J.,* 139:89.

Lipids, Vol. 1, edited by R. Paoletti, G. Porcellati, and G. Jacini. Raven Press, New York © 1976.

Specificity of the Lipase from *Geotrichum candidum*

Robert G. Jensen and Robert E. Pitas

Department of Nutritional Sciences, University of Connecticut, Storrs, Connecticut 06268

Geotrichum candidum is a septate mold, which grows as a white mass on the surfaces of sour cream and cheese. The microorganism was originally investigated by dairy microbiologists because it produced an extracellular lipase that hydrolyzed milk fat. The lipase was found to release largely unsaturated fatty acids from several natural fats (1) and oleic acid (18:1) from synthetic triacylglycerols (2). The latter observations encouraged us to initiate a prolonged study of the specificity of the enzyme. Our approach has generally been to synthesize triacylglycerols of known structure for substrates. The lipase has been the subject of two recent comprehensive reviews (3, 4).

The crude enzyme is readily obtained by dialysis and concentration of the culture medium in which the microorganism is grown (5) and has recently been offered for sale by Worthington Biochemicals, Freehold, New Jersey. An eightfold purification of the enzyme has been achieved by one group with an estimated molecular weight of 37,000 ± 10% (6). Other investigators were able to crystallize the enzyme and obtained a molecular weight of 53,000 to 55,000 (7). This purification resulted in a 40-fold increase in specific activity. The amino acid analysis of the crystalline preparation, revealed that the enzyme did not contain sulfur amino acids. About 7% carbohydrate was found in the crystalline material consisting mainly of mannose, with smaller amounts of xylose, arabinose, and galactose present.

The crude enzyme preparation exhibits optimal activity over broad ranges in pH and temperature, but we routinely assay activity at 37°C and pH 8.1 to 8.2 in Tris buffer. Calcium ions are apparently required for maximum activity (6,8); therefore, the lipase could be a metalloenzyme with calcium involved in the binding site to double bonds. The crude enzyme is exceptionally stable, retaining its activity for at least 9 years at −20°C. If fats containing polyunsaturated fatty acids are to be used as substrates, they should be purified immediately beforehand (9), as oxidation products inhibit activity.

When triacylglycerols of known structure, containing 18:1, were first employed as substrates, the results indicated hydrolysis of 18:1s regardless of whether or not they were primary or secondary esters (2). We confirmed this observation with the data presented in Table 1 (10). These data reveal that the position of oleate within the triacylglycerols did not affect the amount of 18:1 in the free

TABLE 1. *Fatty acid content (M%) of materials from the lipolysis of triacylglycerols by* G. candidum *lipase*

Substrate	Control TG	Residual TG	FFA	DGs	MGs
18:1–16:0–16:0[a]	—	32.6[b]	89.7	1.2	tr
16:0–18:1–18:1	—	64.4[b]	96.1	48.0	tr
16:0–18:1–16:0	—	31.4[b]	94.3	tr	tr
18:1–16:0–18:1	—	66.1[b]	97.7	47.7	tr
18:1–18:1–18:0	—	26.3[b]	82.7	3.2	tr
18:1–18:1–18:1 } 18:2–18:2–18:2 }	58.5[b]	58.4	62.2	56.0	55.0
16:1–18:1–18:1	66.6[b]	67.4	65.2	71.7	68.9
11–18:1–18:1–18:1[a]	74.6[b]	61.6	92.8	37.6	4.4
16:1–6–18:1–6–18:1[a]	33.6[c]	32.4	94.6	0.9	1.4

TG, triacylglycerol; FFA, free fatty acids; DGs, diacylglycerols; MG, monoacylglycerols.

[a] 18:1–16:0–16:0 = *rac* glycerol-1-oleate-2,3 di-palmitate. 11–18:1 = *cis* vaccenate and 6–18:1 = *cis*-petroselinate.

[b] These values are M% oleic acid.

[c] These values are M% palmitoleic acid.

(From Jensen et al., ref. 10.)

fatty acids (FFAs). The rates of digestion of the dioleate substrates were about twice that of monooleate compounds. Thus the extent of hydrolysis was related to the quantities of oleate in the substrates.

Additional results (Table 1) were obtained on the specificity of the lipase by altering the geometry of the substrate. The lipase did not differentiate between 18:1 and 18:2 when the mixed simple triacylglycerols were digested, as all of the hydrolysis products contained the same amount of oleate as the control (see also Table 2). The next comparison was between 18:1 and palmitoleate (16:1); both were hydrolyzed at equimolar rates (Table 1). Shortening of the acyl chain beyond the *cis*-9-double bond did not affect the specificity. We have since found that *cis*-9-14:1 is digested at the same rate as *cis*-9-18:1 *(unpublished data)*.

The enzyme selected *cis*-9-18:1 or *cis*-9-16:1, as compared to two double-bond positional isomers of 18:1 (Table 1). Petroselinate (*cis*-6-18:1) and vaccenate (*cis*-11-18:1) were largely ignored by the lipase as the FFAs were composed of 18:1 or 16:1. Palmitoleate was used in one experiment because it was separable from the 18:1 isomers by the usual gas-liquid chromatographic (GLC) columns, was hydrolyzed by the lipase at the same rate as 18:1, and its use avoided the oxidation procedures that are required to distinguish between 18:1 double bond positional isomers.

Somewhat different results have been obtained by another group of investigators (11) with respect to mixtures of triolein/trilinolein and tripalmitolein/triolein. With both groups of mixed substrates their enzyme preparation exhibited a slight preference for 18:1. These workers noted that 18:2 was hydrolyzed about

10 times more rapidly than petroselinate (*cis*-6-18:1). Linoleic acid (18:3) was also digested, but not as rapidly as 18:1 (11).

With the help of Dr. R. T. Holman of the Hormel Institute who provided us with triacylglycerols containing 18:2, 12:0, 14:0, 16:0, and one each of the double-bond positional isomers of 18:1 from Δ2 to Δ16 (12), we went further into the specificity of the lipase. These were digested individually with *G. candidum* lipase and the compositions of the resulting FFAs are listed in Table 2. It is obvious from these data that *cis*-9-18:1 was the preferred monoene positional isomer. Otherwise 18:2 and the saturates were hydrolyzed. When the *cis*-9-isomer was present, however, both this acid and 18:2 were found in the FFAs in the proportions originally in the triacylglycerols. Although the data are not presented (see ref. 3) the di- and monoacylglycerols from these lipolyses were enriched in the double-bond positional isomers other than *cis*-9-18:1, and as such these fractions might be useful in providing somewhat more concentrated samples of the isomers from partially hydrogenated food fats. We have tested both *cis*-5-14:1 and arachidonic acid (20:4) as substrates and found that compared to 18:1 they are slowly digested *(unpublished data)*.

One of the first questions we asked when we read the paper by Alford et al. (2) was: does the enzyme hydrolyze elaidic acid (*trans* 18:1 or 18:1t) and other trans isomers? To answer the question we synthesized and digested the triacylglycerols listed in Table 3 (13). Only 5.5 M% of 18:1t measured as elaidate by infrared spectrophotometry was found in the FFAs, the remainder being 18:1.

TABLE 2. *Composition of the fatty acids (m%) released by hydrolysis with* G. candidum *lipase from triglycerides containing isomeric* cis-*18:1 acids*

18:1 isomer	Fatty acids				
	12:0	14:0	16:0	18:1	18:2
Δ2	16.5	10.3	14.6	0	58.6
Δ3	9.0	8.6	13.9	3.1	64.8
Δ4	7.8	5.4	7.3	0	79.5
Δ5	8.3	5.0	8.2	2.7	75.8
Δ6	9.4	6.8	14.4	2.0	67.4
Δ7	10.2	5.5	6.8	6.6	70.9
Δ8	8.0	3.7	9.2	4.3	74.8
Δ9	3.5	9.1	17.5	50.6	15.3
Δ10	7.9	4.6	6.6	6.9	74.0
Δ11	7.7	6.7	7.8	4.3	73.5
Δ12	22.8	11.3	15.4	6.7	43.8
Δ13	11.7	8.2	17.6	6.1	56.4
Δ14	17.2	8.6	12.4	5.1	56.7
Δ15	6.5	4.2	10.3	3.9	75.1
Δ16	10.1	8.3	12.0	1.9	67.5

(From Jensen et al., ref. 12.)

TABLE 3. *Fatty acid composition (M%) of acylglycerols and FFAs resulting from lipolysis of triacylglycerols containing* trans *acids and margarine by* G. candidum *lipase*

Substrate and fatty acid composition	Intact TG	Residual TG	FFA	DG	MG
Glycerol-1-elaidate-2,3-dioleate					
18:1 *cis*	70	73.5	94.5	66.0	30.0
18:1 *trans*	30	26.5	5.5	34.0	70.0
Triacylglycerol I					
9,12-*cis, cis*-18:2	47.6	35.2	24.5	29.2	tr
9,12-*trans, trans*-18:2	52.4	64.8	15.5	70.8	90+
Triacylglycerol II[a]					
9,12-*cis, trans*-18:2	46.3	48.5	44.8	49.5	43.5
9,12-*trans, cis*-18:2	53.7	51.5	55.2	51.5	56.5
Margarine					
12:0	tr	tr	tr	tr	5.1
14:0	tr	tr	tr	tr	2.3
16:0	14.0	16.1	7.2	25.2	33.7
18:0	7.2	10.5	3.5	7.4	12.6
18:1	61.2	66.9	74.3	56.8	41.7
18:2	17.6	6.5	15.0	10.6	4.9
trans as elaidic	26.4	32.2	5.5	25.5	16.8

[a] The two isomers were separated by GLC on a column 150 feet × 0.01 inches internal diameter coated with polyphenol ether.
(From Jensen, ref. 3.)

A sample of partially hardened soybean-cottonseed oil margarine was treated similarly, again with very little 18:1t present in the FFAs. Although we were unaware of it at the time, because our GLC column was not capable of separating the compounds, the 18:1 in the di- and monoacylglycerols undoubtedly contained double-bond positional isomers other than *cis*-9. From this and all other substrates containing saturated fatty acids, some of these acids were released.

From the results with *rac* glycerol-1-elaidate-2,3-dioleate shown in Table 3, we were certain that the lipase, which hydrolyzed 18:1t relatively slowly, would behave the same toward *trans, trans*-9,12, and *trans, cis*-9,12-18:2, but would release *cis, trans*-9,12-18:2 thereby separating the cis, trans isomers. Consequently two triacylglycerols were prepared, one containing the cis, cis and trans, trans isomers and the other the cis, trans and trans, cis isomers; these were digested separately (14). The results, in Table 3 at least partially supported our expectations. The cis, cis isomer was released about five times more rapidly than the trans, trans acid. However, much to our surprise, the cis, trans and trans, cis isomers were hydrolyzed at equal rates. The use of internal standards in the GLC analyses indicated that both of these acids were digested at about the same rates as the cis, cis isomer. We have no explanation for this phenomenon.

Our research on the specificity of the lipase has continued with the following results: (a) stearolic acid (octadecynoic acid) in glycerol-1-stearolate-2,3-dioleate,

TABLE 4. *Specificity of the lipase from* G. candidum[a]

Compounds hydrolyzed	Compounds hydrolyzed slowly[b]
cis-9–18:1[c]	4:0–18:0
	trans-9–18:1
cis-9–16:1	*trans, trans*-9,12–18:2
cis-9–14:1	
cis, cis-9,12–18:2	Positional isomers of
	cis 18:1 other than Δ9
cis, trans-9,12–18:2	*cis*-5–14:1
	Arachidonic acid
trans, cis-9,12–18:2	Octadecynoic acid
Palmityloleate	Erucic acid
Cholesteryl oleate	Oleylpalmitate
Linolenic acid	Dilinoleoyl-*sn*-glycerophosphoryl choline[d]

[a] Substrates were triglycerides with obvious exceptions.
[b] Relative to *cis*-9–18:1 or *cis, cis*-9,12–18:2.
[c] Hydrolyzed, regardless of location at positions *sn*-1, 2, or 3, no positional or stereospecificity.
[d] Not hydrolyzed.

dilinoleoyl-*sn*-glycerophosphoryl-choline, and oleylpalmitate were hydrolyzed slowly or not at all and (b) palmityloleate and cholesteryloleate were digested. Erucic acid (*cis*-13-20:1) is also not digested (11, 15).

One of the original reasons for our study of the specificity of *G. candidum* lipase was to determine if the enzyme could be used to analyze the structure of acylglycerols containing 18:1. To this end we synthesized and digested a mixture of the six possible enantiomeric triacid triacylglycerols containing 16:0, 18:0, and 18:1. Stereospecific hydrolysis was not observed (3). Using a fairly complex procedure (3, 4) we were able to determine each isomer in the mixture of six. The method was then applied to the monoene fraction from cocoa butter, and for the first time the individual isomers in a fraction from a natural fat were identified (16).

Data on the specificity of the lipase are summarized in Table 4. The enzyme is the only lipase studied to date which has a true fatty acid specificity; one for *cis*-9- and *cis, cis*-9,12-unsaturation. These acids are hydrolyzed regardless of location within the triacylglycerol. Elaidate (18:1t) and linolelaidate (18:2t) are hydrolyzed slowly compared to the cis acids, but unaccountably both cis, trans and trans, cis-9,12-18:2 are digested. Nevertheless, since *trans* isomers and double-bond isomers other than *cis*-9-18:1 accumulate in the di- and monoacylglycerols resulting from a digestion by the lipase, the enzyme could be used to obtain samples enriched in these acids from partially hydrogenated food fats.

The binding or recognition site for the appropriate double bond(s) is not absolute in that other fatty acids are also digested. Nevertheless, the existence of a recognition site for specifically located double bonds in fatty acids, some distance removed from the active site of the enzyme, is unique and should be further investigated.

SUMMARY

The mold *G. candidum* produces an extracellular lipase that is highly specific for fatty acids containing *cis*-9 and *cis, cis*-9,12-unsaturation. The enzyme hydrolyzes at relatively rapid rates: *cis*-9-18:1, *cis, cis*-9,12-18:2, *cis*-9-16:1, *cis*-9-14:1, palmityloleate, cholesteryloleate, *cis, trans*-9,12-18:2, *trans, cis*-9-18:2, and *cis* 18:3. Compounds hydrolyzed slowly or not at all are: *trans*-9-18:1, *trans, trans*-9,12-18:2, positional double-bond isomers of *cis* 18:1 other than -9, *cis*-5-14:1 octadecynoic acid, erucic acid, arachidonic acid, oleylpalmitate, dilinoleoyl-*sn*-glycerophosphorylcholine, and saturated fatty acids.

The enzyme digests *cis* 18:1 from triacylglycerols regardless of position and is not stereospecific. The positional and trans isomers accumulate in the di- and monoacylglycerols when triacylglycerols containing these fatty acids are digested with *G. candidum* lipase. We suggest that the specificity of the lipase could be used to obtain fractions enriched in these acids from partially hydrogenated food fats.

REFERENCES

1. Alford, J. A., and Pierce, D. A. (1961): *J. Food Sci.,* 26:518.
2. Alford, J. A., Pierce, D. A., and Suggs, F. G. (1964): *J. Lipid Res.,* 5:390.
3. Jensen, R. G. (1974): *Lipids,* 9:149.
4. Brockerhoff, H., and Jensen, R. G. (1974): In: *Lipolytic Enzymes,* p. 140. Academic Press, New York.
5. Alford, J. A., and Smith, J. H. (1965): *J. Am. Oil Chem. Soc.,* 52:1038.
6. Kroll, J., Franzke, C., and Genz, S. (1973): *Pharmazie,* 28:263.
7. Tsujisaka, Y., Iwai, M., and Tomingo, Y. (1973): *Agr. Biol. Chem.,* 37:1457.
8. Tsujisaka, Y., Iwai, M., and Tominaga, Y. (1972): *Proc. IV IFS: Ferment. Technol. Today,* 315.
9. Jensen, R. G., Marks, T. A., Sampugna, J., Quinn, J. G., and Carpenter, D. L. (1966): *Lipids,* 1:451.
10. Marks, T. A., Quinn, J. G., Sampugna, J., and Jensen, R. G., *Lipids,* 3:143.
11. Franzke, C., Kroll, J., and Petzold, R. (1973): *Nahrung,* 17:171.
12. Jensen, R. G., Gordon, D. T., Heimermann, W. H., and Holman, R. T., *Lipids,* 7:738.
13. Jensen, R. G., Sampugna, J., Quinn, J. G., Carpenter, D. L., and Marks, T. A. (1965): *J. Am. Oil Chem. Soc.,* 42:1029.
14. Jensen, R. G., Gordon, D. T., and Scholfield, E. R. (1973): *Lipids,* 8:323.
15. Gurr, M. I., Blades, J., and Appleby, R. S. (1972): *Eur. J. Biochem.,* 29:362.
16. Sampugna, J., and Jensen, R. G. (1969): *Lipids,* 4:444.

Lipids, Vol. 1, edited by R. Paoletti, G. Porcellati, and G. Jacini. Raven Press, New York © 1976.

Reaction Mechanisms and Functions of Plant Lipoxygenases

J. Boldingh

Organic Chemistry Department, State University of Utrecht, The Netherlands

Plant lipoxygenases have long been regarded as an anomalous type of oxygenase in that they were considered to contain no built-in metal. However, Chan (1) reported the detection of one atom of iron per molecule of enzyme isolated from soybeans. Soybeans contain several types of lipoxygenase; the one investigated by Chan was identical to the enzyme already obtained in a crystalline form in 1947 by Theorell et al. (2) (at present this enzyme is known as soybean lipoxygenase-1). Since then several workers (3–6) have confirmed Chan's finding and their attention has been focused on the nature of the active center. Roza and Francke (3) described the extraction of iron by O-phenanthroline under reducing conditions, and in view of the lack of a ferric signal in electron paramagnetic resonance spectroscopy (EPR), they suggested the iron might be present in the ferrous state.

Recently, Pistorius and Axelrod (6) described the removal of iron from the same enzyme by the chelating agent Tiron,[1] and they confirmed the removal of iron by O-phenanthroline in the ferrous state after treatment of the enzyme with a reducing agent. In the latter case, the extraction of the iron is paralleled by a loss in enzyme activity. They concluded that the iron in the enzyme is present in the ferric state.

This conclusion does not fit our observations (7). Evidence for the occurrence of the iron atom in the EPR-silent state in the native soybean lipoxygenase is presented here, but first the various types of reactions known for lipoxygenase-1 under aerobic and anaerobic conditions with linoleic acid as substrate are discussed.

STEREOSPECIFICITY OF LIPOXYGENASE REACTIONS

Lipoxygenase-1 has maximum activity at pH 9.0, whereas for lipoxygenase-2 pH_{opt} is at 6.5 (8). Lipoxygenase-1 converts linoleic acid predominantly into the L-13-hydroperoxide (L-13-hydroperoxy-*cis*-9, *trans*-11-octadecadienoic acid); lipoxygenase-2 gives approximately a 50:50 mixture of L-13- and D-9-hydroperoxide (D-9-hydroperoxy-*trans*-10, *cis*-12-octadecadienoic acid) (9). Figure 1 schematically shows the structures of these compounds.

---

[1] 4,5-Dihydroxy-*m*-benzene disulfonic acid disodium salt.

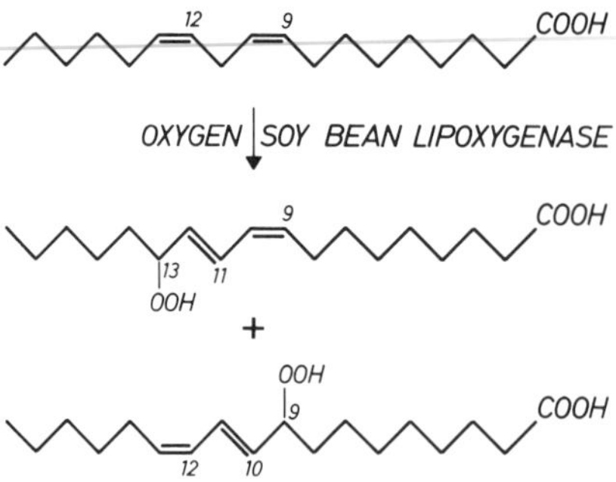

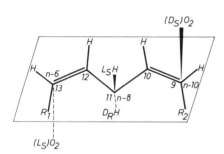

FIG. 1. Conversion of linoleic acid into hydroperoxides by lipoxygenase-1.

From studies with L-11-tritiated linoleic acid Egmond et al. (10) have shown that the hydrogen abstraction proceeds for both conversions antarafacially with respect to oxygen addition in both reactions; abstraction of hydrogen from the 11-pro-L position leads to the formation of L-13-hydroperoxy linoleic acid, whereas abstraction of hydrogen at the 11-pro-D position leads to the formation of D-9-hydroperoxy linoleic acid. This is depicted in Fig. 2 in which the *cis*-1, *cis*-4-pentadiene system of linoleic acid is placed in a planar position.

FIG. 2. The stereochemistry of hydroperoxide formation. R_1:CH_3-$(CH_2)_4$-; R_2:-$(CH_2)_7$-COOH.

PROPERTIES OF SOYBEAN LIPOXYGENASE-1

Activation of the Native Enzyme

The L-13-hydroperoxy linoleic acid is formed with enzyme-1. However, this is not entirely correct. The enzyme can only perform this conversion when the end-product hydroperoxide is present. With thoroughly purified linoleic acid, one observes a lag period, which can last several minutes (11, 12). The lag period (Fig. 3, curve A) can be abolished by the addition of a small amount of hydroperoxy linoleic acid and the effect is specific for the enzymatically prepared L-13-isomer (curve B). In other words the native enzyme must first be activated by the L-13-hydroperoxy linoleic acid, a minute amount of which is already sufficient because of its autocatalytic nature. Once an activated enzyme molecule has been

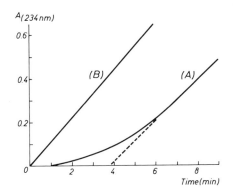

FIG. 3. Lag period in enzymatic conversion of purified linoleic acid **(A)** and the conversion after addition of the L-13-isomer **(B).**

formed in the presence of linoleic acid and oxygen, the system produces its own activating agent.

Anaerobic Lipoxygenase Reaction

The same specificity was found to occur when the now well-known anaerobic lipoxygenase-1 reaction was discovered (13). This reaction proceeds when a mixture of linoleic acid and L-13-hydroperoxy linoleic acid is incubated with

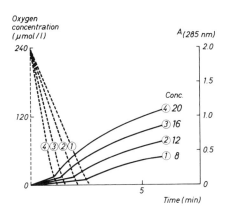

FIG. 4. UV absorptions starting from various initial concentrations (μg/liter) of linoleic acid. The dashed lines represent the corresponding oxygen concentrations.

lipoxygenase-1 under the exclusion of molecular oxygen. Figure 4 shows anaerobiosis obtained by incubating soybean enzyme-1 with linoleic acid with only a limited amount of oxygen in a closed system. At the point of oxygen depletion, the anaerobic reaction starts with the formation of an oxodienoic acid and pentane due to chain fission; the former compound is responsible for the ultraviolet (UV) absorption at 285 nm. The overall chain fission reaction proceeds as follows:

$$CH_3—(CH_2)_4—\underset{\underset{OOH}{|}}{CH}—\overset{t}{CH}=CH—CH\overset{c}{=}CH—(CH_2)_7—COOH$$

$$C_5H_{12}+ \quad \overset{O}{\underset{H}{\diagdown C \diagup}}—\overset{t}{CH}=CH—CH\overset{c/t}{=}CH—(CH_2)_7—COOH$$

It is interesting that concurrent to this reaction dimeric acids are formed of either two linoleic acid molecules or one molecule of linoleic acid and one molecule of hydroperoxy linoleic acid (14). The type of structure of the main dimeric reaction products is

$$—CH_2—\underset{cis}{CH}=CH—CH—\underset{cis}{CH}=CH—CH_2—$$
$$—CH—\underset{trans}{CH}=CH—\underset{cis}{CH}=CH—CH_2—$$

$$—\overset{O}{\overset{\diagup \diagdown}{CH—CH}}—CH—\underset{trans}{CH}=CH—$$
$$\underset{trans}{}$$
$$—CH—\underset{trans}{CH}=CH—\underset{cis}{CH}=CH—$$

$$—\overset{O}{\overset{\diagup \diagdown}{CH—CH}}—\underset{trans}{CH}=CH—CH—$$
$$\underset{trans}{}$$
$$—CH—\underset{trans}{CH}=CH—\underset{cis}{CH}=CH—$$

Again, the anaerobic reaction is specific for the L-13-hydroperoxide; D-9-hydroperoxy linoleic acid does not react under these circumstances. It should be noted that apparently the enzyme is capable of forming linoleic acid radicals even when molecular oxygen is absent. These radicals lead to the formation of the dimers outside the enzyme-substrate complex or in other words they leak out into the solution. We have proven this leakage (11, 15) with 2-methyl-2-nitroso-propan-1-ol as a water-soluble radical trap and linoleic acid that had been deuterated at various positions in the pentadiene system.

$$CH_3—\underset{\underset{NO}{|}}{\overset{\overset{CH_3}{|}}{C}}—\overset{.}{CH_2}—OH + R^. \rightarrow CH_3—\underset{\underset{\underset{R}{|}}{N \overset{.}{—} O}}{\overset{\overset{CH_3}{|}}{C}}—CH_2—OH$$

The EPR spectra produced by the stable nitroxide radicals unambiguously proved that linoleic acid radicals were indeed present in the incubation solution.

Fluorescence Experiments

A third case in which end-product hydroperoxide interacts with the enzyme was discovered by Finazzi Agro et al. (5). They observed fluorescence of soybean lipoxygenase at 328 nm when it is irradiated with ultraviolet light 280 nm (Fig. 5B, curve 1). On removal of oxygen, a slight decrease can be observed (curve 2), but when linoleic acid was added, fluorescence steeply dropped (curve 3). It was soon discovered, in collaboration with Dr. Veldink of our group, that the fluorescence quenching is due to the interaction of enzyme with hydroperoxide formed under aerobic conditions. Indeed the fluorescence quenching can be titrated with enzymatically prepared hydroperoxy linoleic acid and only little more than 1 Equ/mole of enzyme is needed for maximum quenching. Tentatively we then concluded that the low-fluorescing species of the enzyme might represent the activated enzyme. For comparison Fig. 5A gives the weak fluorescence for the heat-inactivated enzyme, which is not influenced by the addition of hydroperoxide.

The Role of Iron in the Enzyme

This conclusion was defined more precisely when we made EPR measurements of lipoxygenase-1 at 15°K at the University of Amsterdam in an attempt to determine the role of iron in the enzyme (7). The native enzyme shows only a moderate signal at $g = 4.3$ and also at $g = 2$ (Fig. 6). The former signal is most likely due to contaminating ferric iron, the signal at $g = 2$ represents very small impurities of manganese and copper (far less than 0.1 Equ). However, when linoleic acid is added aerobically, signals appear at $g = 6.2$, $g = 7.5$, $g = 2$, whereas the signal at $g = 4.3$ increases (Fig. 7).

When, with an excess of linoleic acid, the system is subsequently made anaerobic by depletion of oxygen, the EPR spectrum returns to the original one. Apparently when the hydroperoxide is removed, the iron returns to the EPR-silent (probably ferrous) state, whereas in the presence of hydroperoxide the iron is in the high-spin ferric state. Indeed this proved to be correct. Native lipoxygenase-1 shows the ferric signals when hydroperoxy linoleic acid is added, only little more

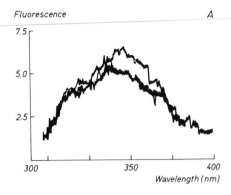

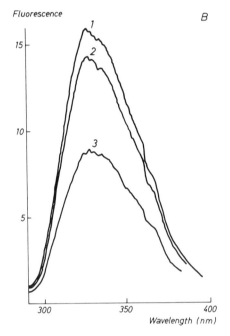

FIG. 5. Fluorescence of soybean lipoxygenase. **(A)** Heat-inactivated enzyme; **(B)** enzyme as such **(1),** after removal of oxygen **(2),** and after addition of linoleic acid **(3).**

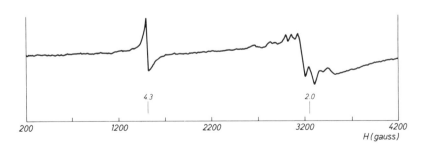

FIG. 6. EPR spectrum of lipoxygenase-1 at 15°K; no linoleic acid added.

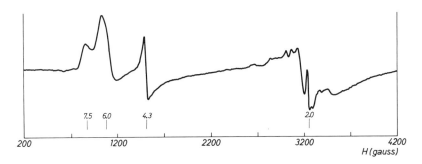

FIG. 7. EPR spectrum of lipoxygenase-1 at 15°K on aerobic addition of linoleic acid.

than 1 Equ with respect to enzyme being needed. Again this holds only for the L-13-hydroperoxy linoleic acid; addition of purified D-9-hydroperoxy linoleic acid has no effect.

It should be noted that during the activation step about 1 Equ L-13-hydroperoxide is decomposed into polar products that we have not yet identified. This decomposition is probably due to the reaction of L-13-hydroperoxide with activated O_2 (superoxide anion?).

Transition of Iron in the Enzyme

Figure 8 gives the proposed scheme for the various transitions of the iron in the active center of soybean lipoxygenase-1 (7). It includes the activation step, the aerobic as well as the anaerobic reaction sequences, and moreover the leakage of linoleic acid radicals. Under aerobic conditions these radicals lead to the formation of small amounts of racemic 13- and 9-hydroperoxy linoleic acids in addition to the bulk of stereospecifically formed L-13-hydroperoxide, whereas in the anaerobic reaction the radicals lead to dimerization reactions. Evidently the knowledge of the anaerobic lipoxygenase reaction has played an essential role in unravelling the nature of the enzyme behavior.

PROPERTIES OF SOYBEAN LIPOXYGENASE-2

Soybean lipoxygenase-2, which works best at pH 6.5, tends to be more leaky with respect to linoleic acid radicals than lipoxygenase-1 and, as a consequence, needs more reactivation during the aerobic reaction. We believe that this is the cause of the co-oxidation reactions, which have been reported in the literature by, for instance, Arens et al. (16) on carotenoids and Teng and Smith (17) on cholesterol. The enzyme is moreover less stable than lipoxygenase-1 and loses its iron more readily, as has also been published lately (6).

It is peculiar that, depending on the duration of swelling with water, the soybean can contain amounts of lipoxygenase-2 varying from a large proportion

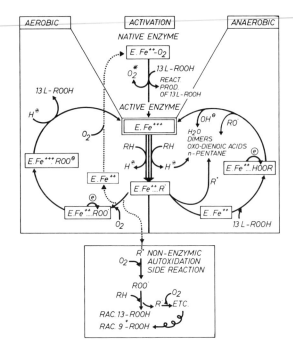

FIG. 8. Scheme of the various transitions of the iron in the active center of soybean lipoxygenase-1.

to practically nil. Some details of this phenomenon have been discussed by Veldink, Slappendel, Vliegenthart, and Boldingh *(This Volume),* whereas Roza *(This Volume)* has reported on yet another lipoxygenase fraction of soybeans, which acts on fatty acid esters and forms cyclic hydroperoxy peroxides from the methyl ester of α-linolenic acid. In other sources of plant lipoxygenase, e.g., in other legume seeds (18), or maize (19), and also potatoes (20), the lipoxygenase is predominantly of type 2.

PHYSIOLOGICAL FUNCTIONS OF LIPOXYGENASE

It has been suggested on several occasions (e.g., by Nichols, ref. 21) that lipoxygenase is involved in cutin formation in plant epidermis cells. Cutin, when partly saponified, yields fragments that contain interesterified oxygenated C_{18} fatty acid chains. These elements of plant cutin suggest the involvement of lipoxygenase in cutin biosynthesis, but thus far no direct evidence has been produced.

In our opinion the anaerobic reaction might have functional significance, e.g., in the dormant soybean seeds, in that the enzyme protects the seeds by keeping the oxygen concentration at a low level, whereas it eliminates autoxidatively formed hydroperoxide at the same time along the anaerobic pathway (13). In this connection it is interesting to mention that Pattee (22) observed a release of pentane from maturing peanuts; the maximum release coincides with a maximum of total lipoxygenase activity in the seeds. It might even be possible that pentane itself exerts a specific function at the biomembrane level. Much work has still to be done in order to obtain relevant information on the physiologic function of these enzymes; at present we can only put forward rather speculative hypotheses.

SUMMARY

The isolation and characterization of various types of plant lipoxygenases are discussed. Particular attention is paid to the purification of soybean lipoxygenase-1, its modes of action—both in the aerobic and anaerobic reaction—the induction period and the selective role of L-13-hydroperoxy linoleic acid (end-product) in enzyme activation.

These aspects are discussed in conjunction with fluorescence quenching by the end-product and also with regard to recent findings in EPR studies that show that the state of iron in the active center of the enzyme is selectively dependent on the presence of L-13-hydroperoxy linoleic acid. Finally, some suggestions are made about possible physiologic functions of lipoxygenases *in vivo*.

ACKNOWLEDGMENT

The work described is a result of the study carried out by a team at the State University of Utrecht to which also participate J. F. G. Vliegenthart, G. A. Veldink, G. J. Garssen, M. R. Egmond, J. J. M. C. de Groot, J. Verhagen, and L. J. M. Spaapen.

REFERENCES

1. Chan, H. W. S. (1972): Communication to the 11th World Congress of the International Society for Fat Research, Göteborg, Sweden.
2. Theorell, H., Holman, R. T., and Åkeson, A. (1947): *Acta Chim. Scand.,* 1:571.
3. Roza, M., and Francke, A. (1973): *Biochim. Biophys. Acta,* 327:24.
4. Chan, H. W. S. (1973): *Biochim. Biophys. Acta,* 327:32.
5. Finazzi-Agrò, A., Avigliano, L., Veldink, G. A., Vliegenthart, J. F. G., and Boldingh, J. (1973): *Biochim. Biophys. Acta,* 326:462.
6. Pistorius, E., and Axelrod, B. (1974): *J. Biol. Chem.,* 249:3183.
7. De Groot, J. J. M. C., Veldink, G. A., Vliegenthart, J. F. G., Boldingh, J., Wever, R., and Van Gelder, B. F. (1975): *Biochim. Biophys. Acta, in press.*
8. Christopher, J., Pistorius, E., and Axelrod, B. (1970): *Biochim. Biophys. Acta,* 198:12.
9. Christopher, J., and Axelrod, B. (1971): *Biochem. Biophys. Res. Commun.,* 44:731.
10. Egmond, M. R., Vliegenthart, J. F. G., and Boldingh, J. (1972): *Biochem. Biophys. Res. Commun.,* 48:1055.
11. Garssen, G. J. (1972): Thesis, University of Utrecht, The Netherlands.
12. Smith, W. L., and Lands, W. E. M. (1972): *J. Biol. Chem.,* 247:1038.
13. Garssen, G. J., Vliegenthart, J. F. G., and Boldingh, J. (1971): *Biochem. J.,* 122:327.
14. Garssen, G. J., Vliegenthart, J. F. G., and Boldingh, J. (1972): *Biochem. J.,* 130:435.
15. De Groot, J. J. M. C., Garssen, G. J., Vliegenthart, J. F. G., and Boldingh, J. (1973): *Biochim. Biophys. Acta,* 326:279.

16. Arens, D., Seilmeier, W., Weher, F., Kloos, G., and Grosch, W. (1973): *Biochim. Biophys. Acta,* 327:295.
17. Teng, J. I., and Smith, L. L. (1973): *J. Am. Chem. Soc.,* 95:4060.
18. Eriksson, C. E., and Svensson, S. G. (1970): *Biochim. Biophys. Acta,* 198:449.
19. Veldink, G. A., Garssen, G. J., Vliegenthart, J. F. G., and Boldingh, J. (1972): *Biochem. Biophys. Res. Commun.,* 47:22.
20. Galliard, T., and Phillips, D. R. (1971): *Biochem. J.,* 124:431.
21. Hitchcock, C., and Nichols, B. W. (1971): In: *Plant Lipid Biochemistry,* p. 191. Academic Press, New York.
22. Pattee, H. E., Singleton, J. A., Johns, E. B., and Mullin, B. C. (1970): *J. Agr. Food Chem.,* 18:353.

Lipids, Vol. 1, edited by R. Paoletti, G. Porcellati, and G. Jacini. Raven Press, New York © 1976.

Enzyme-Generated Free Radicals and Singlet Oxygen as Promoters of Lipid Peroxidation in Cell Membranes

Paul B. McCay, Kuo-Lan Fong, Margaret King, Edward Lai, Charles Weddle, Lee Poyer, and K. Roger Hornbrook*

*Biomembrane Research Laboratory, Oklahoma Medical Research Foundation, Oklahoma City, Oklahoma 73104, and *Department of Pharmacology, College of Medicine, University of Oklahoma Health Sciences Center, Oklahoma City, Oklahoma 73190*

Although lipid peroxidation has been the subject of many investigations, its precise role (if any) in animal cells is not known. The requirement for a dietary free-radical scavenger to maintain integrity of certain tissues in animals together with the accumulation of ceroid pigments in animal cells as they age suggests that lipid peroxidation may be a significant process in living organisms. Most of the studies have explored the possibility that autoxidation of endogenous lipids in cells may occur through nonenzymatic mechanisms caused by substances such as ascorbate, glutathione, and cysteine, and it is possible that such substances do promote some lipid peroxidation *in vivo*. Tappel has shown that heme groups are also capable of promoting lipid peroxidation (1). We believed that the endogenous concentration of oxygen is probably too low in most tissues to support any significant amount of autocatalytic lipid peroxidation. However, our studies have led us to conclude that electron transporting systems containing flavin enzymes would be the most likely promotors of lipid peroxidation in the intact cell (2). We have demonstrated that the activity of at least three different enzymes (two of which are known to be flavoproteins), can initiate lipid peroxidation in biologic membranes, because they promote the formation of highly reactive free radicals and apparently singlet oxygen as well. It was observed that a low concentration of Fe^{3+} was required for this activity and that the cytosol of most tissues contains sufficient iron to promote the activity. The two flavoenzymes we have studied most thoroughly are purified liver microsomal NADPH-Cytochrome P_{450} reductase, and purified xanthine oxidase. Both of these enzymes have been reported to produce superoxide anion (3, 4). In part of these studies, we used liver lysosomes as a biologic membrane system to test for free-radical damage as a result of exposure of these particles to the activity of these flavin enzymes. Lysosomal membranes contain unsaturated lipids susceptible to peroxidation, the extent of which can be assayed by measurement of malondialdehyde, which is formed by the breakdown of fatty acid hydroperoxides.

MATERIALS AND METHODS

Animals

Young adult albino male rats (250 to 350 g) of the Sprague-Dawley strain were used in these studies.

Materials

p-Chloromercuribenzoic acid, D-glucose-6-phosphate disodium salt, NADP, glucose-6-phosphate dehydrogenase, xanthine oxidase, Triton X-100, cytochrome c Type III, hyaluronidase (ovine, Type III), and NADPH were obtained from the Sigma Chemical Company. Santoquin (ethoxyquin), 2-thiobarbituric acid and ethylenediaminetetraacetate (EDTA) were from Eastman Organic Chemicals (Rochester, New York). *p*-Nitrophenol, *p*-nitrophenyl phosphate, *trans*-1,2-dibenzoylethylene, and *o*-dibenzoylbenzene were purchased from the Aldrich Chemical Company. *cis*-Dibenzoylethylene was prepared by our laboratory. Rose bengal was obtained from the Koch-Light Laboratories. Sucrose and mannitol were supplied by the Fisher Scientific Company. Triton WR 1339 was obtained from the Ruger Chemical Company (Irvington, New Jersey). Carbon tetrachloride came from the J. T. Baker Company (Phillipsburg, New Jersey). Trypan blue stain was purchased from the Curtin Scientific Company (Tulsa, Oklahoma) and Collagenase (CLS II) from the Worthington Biochemical Corporation (Freehold, New Jersey). Other chemicals and solvents were of reagent grade quality.

Methods

Preparation of subcellular particles. Liver lysosomes were prepared as described by Fong et al. (2). Liver microsomes were prepared by the procedure described by May and McCay (5). Mitochondria and the plasma membrane-nuclei fraction were prepared as described by Pfeifer and McCay (6).

Isolated rat liver cells. Rat liver parenchymal cells were isolated by minor modifications of an *in situ* enzymatic procedure described by Berry and Friend (7). The liver was perfused without recirculation with 60 ml of Hanks buffer solution (Ca^{2+} and glucose-free) gassed with $O_2:CO_2$ (95:5). Then hyaluronidase (0.05%) and collagenase (0.02%) were added to the perfusate buffer solution, and it was recirculated through the liver at a flow rate of 20 to 25 ml/min for 20 to 30 min. The liver was excised and minced, and fresh gassed-perfusion medium was added. Calcium chloride was added to a final concentration of 1.0 mM. Then the cells were filtered and washed. The hepatocytes were counted with a Bright-Line hemocytometer. The integrity of the plasma membranes of these cells was analyzed with Trypan blue stain. Those cell preparations in which more than 85% of the cells excluded the dye were used for these experiments.

Incubation systems: Lysosomal studies. The incubation systems were as described in Fong et al. (2). *Singlet oxygen studies.* Photochemical generation of singlet oxygen with rose bengal was carried out according to the procedure of Porter and Ingraham (8). Singlet oxygen generation by microsomal NADPH oxidase activity was assayed by the addition of diphenylfuran (DPF) to the reaction system as described by May and McCay (5).

Isolated liver cells were incubated in Hanks medium containing additional 10-mM sodium phosphate solution, pH 7.4. The liver cells used for the time course studies were incubated in test tubes. When rat hepatocytes were incubated for the studies of lipid content of subcelluar membrane fractions, the liver cells were incubated in Erlenmeyer flasks, which were gassed with pure O_2 and subsequently stoppered. Carbon tetrachloride was added to the center wells of these flasks and allowed to diffuse into liver cell suspension prior to the addition of NADPH. After incubation the hepatocytes were homogenized and fractionated into nuclei-plasma membranes, microsomes, and mitochondria by differential centrifugation.

Assays. Malondialdehyde formation was measured as described in the report by Fong et al. (2). The fatty acid composition of the various liver fractions was performed as described by May and McCay (5). The assay for superoxide anion generation by xanthine oxidase activity was as described by McCord and Fridovich (4). Protein content of preparations was measured by the method of Lowry et al. (9).

Synthesis of cis-Dibenzoylethylene Standard. Fresh *cis*-dibenzoylethylene standard was prepared by the method of Lutz and Wilder (10).

RESULTS

Free-radical generation during flavin enzyme activity in the presence of Fe^{3+}

As indicated in earlier studies, the activity of certain flavin enzymes is associated with production of O_2^- (2), which, however, is completely unreactive with tissue lipids unless Fe^{3+} is present in the system. The inclusion of iron in the systems promoted the formation of highly reactive hydroxyl radicals that readily attacked lipids. Table 1 shows typical results obtained when liver lysosomes were incubated with the purified flavin enzyme, NADPH-Cytochrome P_{450} reductase. Very little release of lysosomal hydrolases occurred when the particles were exposed to the enzyme in the presence of its substrate, NADPH. In addition, there was very little lipid peroxidation in the system. However, when Fe^{3+} was added to the system in the form of an adenosine diphosphate (ADP) complex, both hydrolase release and malondialdehyde formation was stimulated indicating peroxidative attack on the lysosomal membrane had occurred. This attack was substantially prevented by including a free-radical scavenger in the system (Table 1).

Figure 1 shows the results of studies with another purified flavin enzyme,

TABLE 1. *Lysosomal lysis caused by cytochrome P_{450} reductase activity; a peroxidative process requiring chelated iron*

Additions to incubation medium	Free acid phosphatase released (% of total)	Malondialdehyde formation (nm moles/ml reaction system)
LYS	8.99	2.24
LYS + CYT P_{450} RED	11.49	3.58
LYS + CYT P_{450} RED + NADPH	12.99	4.93
LYS + CYT P_{450} RED + ADP-Fe^{3+} + NADPH	42.99	18.14
LYS + CYT P_{450} RED + ADP-Fe^{3+}	18.99	9.41
LYS + ADP-Fe^{3+} + NADPH	19.41	8.51
LYS + ADP-Fe^{3+}	12.99	5.82
LYS + CYT P_{450} RED + ADP-Fe^{3+} + NADPH + santoquin	19.49	5.71

The basic medium contained 0.7-M sucrose in 0.015-M Tris, pH 7.4. Additions were made as indicated in the table. Lysosomes (Lys), 0.3–0.5 mg-protein/ml system; NADPH, 0.3 mM; ADP, 2.0 mM; Fe^{3+}, 0.12 mM; purified NADPH-dependent cytochrome P_{450} reductase (Cyt P_{450} Red), 0.066-mg protein/ml system; Santoquin, 1.0 x 10^{-3} M. Incubations were carried out at 37°C for 30 min and followed by addition of 1.0 mM Mn^{2+} to stop further action on lysosomes. Assay of acid phosphatase released was performed according to the method of Fukuzawa et al. (11). Assay of malondialdehyde formation is described under "Methods."

xanthine oxidase. This enzyme is known to form the superoxide anion radical, the production of which can be assayed by measuring cytochrome c reduction: Cytochrome $c^{3+} + O_2^- \rightarrow$ Cytochrome $c^{2+} + O_2$ (4). Figure 1 shows that xanthine oxidase-catalyzed reduction of cytochrome c (via (O_2^-)) is markedly inhibited by the addition of complexed Fe^{3+} and, in fact, that part of the cytochrome c that had been reduced starts to undergo reoxidation. Xanthine oxidase activity itself was not inhibited by the iron addition as shown by the fact that uric acid formation was not affected by addition of Fe^{3+} (Fig. 1). The reoxidation of the reduced cytochrome c caused by ADP-Fe^{3+} was not caused by reduction of Fe^{3+} by the cytochrome but rather is caused by generation of an oxidizing free radical. Figure 2 shows evidence that the radical is HO·. In this experiment the reoxidation of reduced cytochrome c promoted by xanthine oxidase activity in the presence and absence of Fe^{3+} was studied. The data show that free radical scavengers inhibit the reoxidation of reduced cytochrome c. The effectiveness of two of these scavengers (mannitol and ethanol) indicate that the free radical involved is HO· (12). The scavengers are effective whether added at the beginning of the reaction or at 2 min after initiating the reaction. These studies clearly

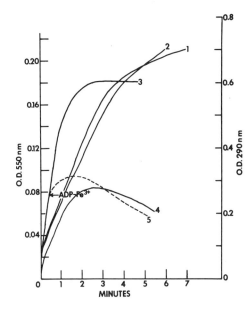

FIG. 1. The effect of ADP-Fe^{3+} on the production of uric acid and the production of superoxide anion generated by the xanthine oxidase activity. Uric acid formation was determined by measuring the increase in absorbancy at 290 nm. Standard reaction mixtures for uric acid production consisted of 5×10^{-5}M xanthine and 0.05-mg protein/ml system xanthine oxidase in a final volume of 1.5 ml buffered at pH 7.8 with 0.05-M potassium phosphate. Additions to the standard reaction mixture for uric acid production were curve **1,** none; curve **2,** 2.0-mM ADP and 6.6×10^{-5}M Fe^{3+}. Cytochrome c reduction was assayed by the method of McCord and Fridovich (4). Standard reaction mixture for cytochrome c reduction consisted of 5×10^{-5}M xanthine, 0.05-mg protein/ml system xanthine oxidase, and 1.3×10^{-5}M ferricytochrome c in a final volume of 1.5 ml buffered at pH 7.8 with 0.05-M potassium phosphate. Additions to the standard reaction mixture for cytochrome c reduction were curve **3,** none; curve **4,** 2.0-mM ADP and 6.6×10^{-5}M Fe^{3+} at 0 time; curve **5,** 2.0-mM ADP and 6.6×10^{-5}M Fe^{3+} at 30 sec after the reaction was initiated.

indicate the production of hydroxyl free radicals by the activity of certain flavin enzymes that produce superoxide anion; however, Fe^{3+} is required for efficient conversion of O_2^{-} to HO· radicals.

Singlet Oxygen Generation by Microsomal NADPH Oxidase Activity

DPF reacts with singlet oxygen to form dibenzoylethylene (8). Washed microsomes, NADPH (or an NADPH-generating system), DPF, and 0.012-mM Fe^{3+} (in the form of an ADP complex) were incubated for 45 min and then the total lipids were extracted with chloroformmethanol (2:1), and the lipid material recovered, following the procedure of Folch et al. (13). This procedure not only extracts DPF but dibenzoylethylene as well. When the lipid extract was analyzed by thin-layer chromatography (TLC), dibenzoylethylene could be detected on the

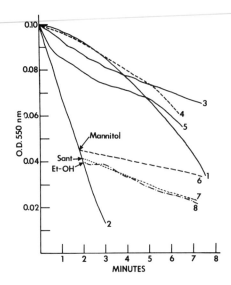

FIG. 2. Effect of free radical scavengers on the oxidation of reduced cytochrome c by xanthine oxidase. Ferrocytochrome c was prepared by reduction of ferricytochrome c with ascorbate followed by dialysis at 0°C and under pure nitrogen, against 500 volumes of 0.05-M potassium phosphate at pH 7.8 for 24 hr. Oxidation of ferrocytochrome c was determined by measuring the decrease in absorbancy at 550 nm. The standard reaction mixture for oxidation of ferrocytochrome c consisted of 6.6×10^{-5}M xanthine, 0.05-mg protein/ml system xanthine oxidase, 2.0-mM ADP, 6.6×10^{-5}M Fe^{3+}, and 1.3×10^{-5}M ferrocytochrome c in a final volume of 1.5-ml buffered at pH 7.8 with 0.05-M potassium phosphate. Additions to the standard reaction mixture were curve 1, ADP-Fe^{3+} was omitted; curve 2, none; curve 3, 3.3×10^{-2}M mannitol at 0 time; curve 4, 0.66 M ethanol at 0 time; curve 5, 1.0×10^{-3} M Santoquin (Sant) at zero time; curve 6, 3.3×10^{-2} M mannitol at 2 min after the reaction was initiated.

plate (Fig. 3). Not only did the material have the same R_f as authentic dibenzoylethylene, but it also had the same infrared spectrum.

That the production of this singlet oxygen reaction product of DPF was due to microsomal NADPH oxidase activity was determined in several ways. First, as seen in Fig. 3, omitting the substrate NADPH (or the NADPH-generating system) resulted in no formation of dibenzoylethylene. Also prior heating of the microsomes at 70°C for 2 min before addition to the reaction system completely abolished dibenzoylethylene production. Addition of sulfhydryl reagent such as p-chloromercuribenzoate also abolished the formation of dibenzoylethylene. Assuming that dibenzoylethylene formation is indicative of singlet oxygen formation (8), microsomal NADPH oxidase activity generates singlet oxygen. Iron is essential for the formation of singlet oxygen and if it is not added to the washed microsomes, very little dibenzoylethylene is formed. That singlet oxygen formation in this microsomal system may play a role in the lipid peroxidation that accompanies NADPH oxidation is supported by the data in Table 2. The addition of increasing concentrations of DPF to the reaction system results in a more intense formation of dibenzoylethylene as judged by TLC analysis. At the same

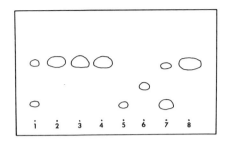

FIG. 3. TLC of lipids extracted from liver microsomes after a 45-min incubation in a standard enzyme incubation system **(1–6)** or under light 20 min in a photodynamic system **(7, 8)**. The standard enzyme system consisted of 0.1 mg of microsomal suspension per milliliter of reaction system (the equivalent of about 1 mg of protein), and 0.012-mM $FeCl_3$ complexed with 4-mM ADP, 0.3-mM NADPH, and in 0.05-M Tris-HCl buffer, pH 7.5 with 1-mM DPF in acetone added to each system: (1) complete enzyme system, (2) without NADPH, (3) without ADP-Fe^{3+}, (4) DPF standard, (5) *cis*-dibenzoylethylene standard, (6) *trans*-dibenzoylethylene standard. The photodynamic system consisted of 0.1 mg microsomal suspension per milliliter of incubation system and 15-μM rose bengal in $D_2O:H_2O$ (1:1) and containing 1-mM DPF (7) complete photodynamic system (8) without rose bengal.

time, malondialdehyde formation is inhibited inversely, suggesting that DPF is competing with the lipid for singlet oxygen. Singlet oxygen is known to cause lipid peroxidation (14). It was not possible to completely inhibit malondialdehyde production with DPF, which indicates that malondialdehyde production, which is insensitive to DPF, may be due to hydroxyl radical formation, which, as described earlier, also apparently occurs in this system.

TABLE 2. *Effect of DPF on lipid peroxidation in liver microsomes catalyzed by NADPH oxidation*

DPF added to system (final conc.) (mM)	Malondialdehyde formed/mg microsomal protein (nmoles)	Dibenzoylethylene formation (thin-layer plate analysis)
0	8.00	—
0.05	5.76	+
0.125	5.03	+
0.25	4.78	++
0.375	4.72	++
0.5	4.38	+++
0.75	4.16	+++
1.00	3.79	++++

The incubation system consisted of 0.1 ml of microsomal suspension per ml of reaction system (the equivalent of about 1 mg of protein) 0.012-mM $FeCl_3$ complexed with 4-mM ADP, 0.3-mM NADPH, and in 0.05-M Tris-HCl buffer at pH 7.5. Assay for malondialdehyde formation was by the thiobarbituric acid assay as described by Fong et al. (2). Thin-layer analysis of dibenzoylethylene formation was as described in Fig. 3.

Evidence that Lipid Peroxidation Occurs in the Intact Liver Cell

When isolated liver cells from normal rats were incubated in a buffered supporting medium, malondialdehyde formation was observed in the system (Table 3), and the amount formed increased with time. When rats had been pretreated with phenobarbital (75 mg/kg i.p. daily for 5 days), the liver cells from these animals exhibited an increased lipid peroxidation indicating that much of this material must be formed from lipids in the endoplasmic reticulum, since this membranous organelle increased in amount two- to threefold by phenobarbital treatment.

To ascertain if lipid peroxidation in intact liver cells could be increased by factors previously used in studies with microsomal preparations, the cells were incubated with an additional source of NADPH and Fe^{3+}. Figure 4 shows that malondialdehyde formation was stimulated by the addition of NADPH and Fe^{3+}. On analyzing the fatty acid composition of cells immediately after incubation, however, the polyunsaturated fatty acids of the various membrane fractions indicated that the peroxidation caused by the addition of NADPH and Fe^{3+} occurred only in the plasma membrane (Table 4). This finding suggests that there is also an enzyme system in the plasma membrane that can oxidize NADPH and initiate radical formation in the presence of iron.

Isolated rat liver cells were also incubated with CCl_4, which is known to be metabolized by an NADPH-requiring system in liver microsomes and causes lipid peroxidation in these particles (15). Figure 5 shows that CCl_4 significantly stimulated malondialdehyde formation by isolated liver cells. Analyses of fatty acids in subcellular fractions of these cells showed a loss of approximately 50% of the phospholipids in the endoplasmic reticulum and some loss in the nuclei-plasma membrane fraction (Table 4). Electron micrographs showed that the

TABLE 3. *Lipid peroxidation in isolated liver cells from normal and phenobarbital-induced rats*

	Thiobarbituric acid reaction (O.D.$_{532nm}$) values obtained after incubating 10^8 cells	
	5 min	30 min
Normal cells	0.43	2.64
PB-induced cells	0.92	4.91

The liver cells were suspended in Hanks buffer containing an additional 10-mM sodium phosphate, pH 7.4. A 1.0-ml suspension of hepatocytes was preincubated with shaking in a 37°C water bath for 17 min. After incubating the liver cells an additional 5 and 30 min, 0.5-ml 35% trichloroacetic acid was added to the incubation systems. The thiobarbituric acid analysis of malondialdehyde production is described in "Methods." The O.D.$_{532nm}$ of cells preincubated for 17 min was subtracted from those values of O.D.$_{532nm}$ obtained from liver cells incubated for the 5 and 30 min time periods. Data are results of five to six experiments.

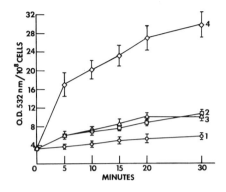

FIG. 4. The incubation conditions and procedures are described in Table 3. After 15 min of preincubation (zero time), the liver cells isolated from normal rats were incubated with various substrates: curve 1, cells only; curve 2, NADPH: 0.6 mM; curve 3, ADP-Fe^{3+}: 0.4 mM; curve 4, NADPH: 0.6 mM and ADP-Fe^{3+}: 0.4 mM. The data plotted is an average of five to six experiments ± SEM.

endoplasmic reticulum was severely disrupted in these cells as a result of exposure to CCl_4. This effect was apparently not due to the solvent nature of CCl_4 as indicated by the lack of an effect on the cells when exposed to CH_2Cl_2. These studies give strong support to the hypothesis that the damage caused by the metabolism of CCl_4 in liver tissue is due to peroxidation of lipids in the endoplasmic reticulum. It was of interest that mitochondrial lipids were not affected by any condition used. This correlates with known data that indicate the structure

TABLE 4. *Fatty acid composition of subfractions of isolated liver cells incubated in systems promoting lipid peroxidation*

Incubation systems	Fatty acid content (mg/10^8 cells)					
	16:0	18:0	18:1	18:2	20:4	22:6
Microsomes						
Liver cells only	0.69	1.11	0.34	0.59	1.07	0.18
Liver cells + NADPH + ADP-Fe^{3+}	0.87	1.41	0.43	0.77	1.27	0.21
Liver cells + CCl_4	0.44	0.70	0.25	0.39	0.58	0.11
Nuclei-plasma membranes						
Liver cells only	2.41	3.10	1.53	2.91	2.77	0.57
Liver cells + NADPH + ADP-Fe^{3+}	2.06	2.54	1.42	2.48	1.99	0.38
Liver cells + CCl_4	2.32	2.63	1.64	2.60	1.81	0.36

The basal incubation medium was Hanks buffer + 10-mM phosphate. NADPH: 0.6 mM, ADP-Fe^{3+}: 0.40mM, CCl_4: saturated by gas phase diffusions were added as described in "Methods." Data are results of four experiments. Underlined values are significant from those found for liver cells only (PL 0.05).

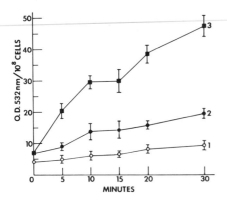

FIG. 5. Liver cells isolated from phenobarbital pretreated rats were preincubated with shaking in a 37°C water bath for 2 min. Then carbon tetrachloride was added to a final concentration of 10 μl/ml, except for curve 1, and the cells were incubated 15 min. Curve **1**, cells only; curve **2**, 10 μl/ml CCl$_4$; curve **3**, 10 μl/ml CCl$_4$ and NADPH: 0.6 mM.

of these particles, as determined by electron microscopy, are unaffected by CCl$_4$ toxicity (16).

DISCUSSION

The three sets of studies described above provide evidence that metabolic activities of some enzymes in liver cells may be accompanied by the formation of highly reactive free radicals and singlet oxygen, which are capable of causing peroxidative damage to membranous structures of the cell. An earlier study of this phenomenon from this laboratory had demonstrated that superoxide anion was very likely the initial radical produced by the enzymes that promote lipid peroxidation in liver microsomes and lysosomes but that this radical did not attack the membrane lipids. Rather, superoxide appears to undergo both a dismutation to hydrogen peroxide and a reaction with iron to promote the formation of hydroxyl radicals (17). It now also appears that singlet oxygen, which is formed by another type of dismutation of superoxide (18) is produced when iron is present in the system. Flavin enzymes, especially those that may contain or bind iron, may be especially capable of catalyzing such reactions. The studies also indicated that such processes may occur in intact, functioning liver cells alone because they sustain a low level of on-going peroxidative activity that does not affect the unsaturated fatty acid composition of its membranous parts. The inability to detect changes in fatty acid composition could be due to homeostatic mechanisms that repair damaged phospholipids or to the small amounts of altered lipids formed. However, when radical generation is augmented by exogenous factors—and there are many such factors—markedly enhanced peroxidative attack is observed on the unsaturated fatty acids of the endoplasmic reticulum and apparently in the plasma membranes of intact cells. This study with isolated liver cells may be another indication that lipid peroxidation is a significant factor in some metabolic processes. Lipid peroxidation has been associated with the phagocytic process in intact leukocytes (19). Because of the growing number of enzyme systems that produce superoxide anion, it would appear that lipid peroxidation promoted by the hydroxyl radical and singlet oxygen mechanisms described in this report may be a general phenomenon in animal tissues and is probably held within manageable limits normally by a combination of the action of three factors: (a) Superoxide dismutase, (b) dietary free radical scavengers, and (c) homeostatic replacement of altered lipids. Various evidence suggests that reduction of any one of these factors results in some tissues being damaged by lipid peroxidation.

SUMMARY

Although indirect evidence has existed for sometime that lipid peroxidation occurs in animal tissues, the molecular mechanisms by which the process occurs are not well defined. Recently we have demonstrated that certain constitutive enzymes in mammalian tissues can promote lipid peroxidation in membranous subcellular organelles apparently by generating hydroxyl free radicals during the activity of these enzymes (2). The studies indicated that the hydroxyl radicals may be formed secondarily to the production of superoxide anion radicals by enzymic reactions. Two enzyme systems that can promote lipid alterations in biologic membranes are flavin enzymes (xanthine oxidase and NADPH-cytochrome P_{450} reductase) and are known to produce superoxide during their activity. It was conclusively shown, however, that the superoxide anion itself does not attack polyunsaturated lipids directly. We have recently obtained evidence that (a) such a mechanism for lipoperoxidation can operate in intact cells; (b) low levels of lipoperoxidation may be a continuous process in cells containing such enzymes; and (c) singlet oxygen associated with enzyme activity may be involved in lipid peroxidation.

ACKNOWLEDGMENTS

This work was supported in part by U.S. Public Health Service Grants AM-06978, AM-08397, and AM-16551.

The authors wish to thank Drs. Bernard Keele and Hara Misra for providing superoxide dismutase, Donald Gibson for technical aid, and Wanda Honeycutt and Rhonda Goodenow for assistance in the preparation of this report.

REFERENCES

1. Tappel, A. L. (1962): In: *Hematin Compounds and Lipoxidase as Biocatalysts, Symposium on Foods: Lipids and Their Oxidation,* edited by H. W. Schultz, pp. 122. Avi Publishing, Westport, Conn.
2. Fong, K. L., McCay, P. B., Poyer, J. L., Keele, B. B., and Misra, H. (1973): *J. Biol. Chem.,* 248:7792.
3. Strobel, H. W., and Coon, M. J. (1971): *J. Biol. Chem.,* 246:7826.
4. McCord, J. M., and Fridovich, I. (1969): *J. Biol. Chem.,* 244:6049.
5. May, H. E., and McCay, P. B. (1968): *J. Biol. Chem.,* 243:2288.
6. Pfeifer, P. M., and McCay, P. B. (1972): *J. Biol. Chem.,* 247:6763.

7. Berry, M. N., and Friend, D. S. (1969): *J. Cell Biol.,* 43:506.
8. Porter, David J. T., and Ingraham, L. L. (1974): *Biochim. Biophys. Acta,* 334:97.
9. Lowry, H. O., Rosebrough, N. J., Farr, A. C., and Randall, R. S. (1951): *J. Biol. Chem.,* 193:265.
10. Lutz, R., and Wilder, D. (1934): *J. Am. Chem. Soc.,* 56:978.
11. Fukuzawa, K., Suzuki, Y., and Uchiyama, M. (1971): *Biochem. Pharmacol.,* 20:279.
12. Neta, P., and Dorfman, L. M. (1968): *Adv. Chem.,* 81:222.
13. Folch, J., Lees, M., and Sloane-Stanley, G. H. (1957): *J. Biol. Chem.,* 226:497.
14. Rawls, H. C., and van Santen, P. J. (1970) *Anal. N.Y. Acad. Sci.,* 171:135.
15. Recknagel, R. O., and Glende, E. A., (1973): *Crit. Rev. Toxicol.,* 2:263.
16. Krishnan, N., and Stenger, R. J. (1966): *Am. J. Pathol.,* 49:239.
17. Haber, F., and Weiss, J. (1934): *Proc. Roy. Soc. Edinburgh,* A147:332.
18. Khan, A. U. (1970): *Science,* 168:476.
19. Stossel, T. P., Mason, R. J., and Smith, A. L. (1974): *J. Clin. Invest.,* 54:638.

Lipids, Vol. 1, edited by R. Paoletti, G. Porcellati, and G. Jacini. Raven Press, New York © 1976.

Lipid–Sterol Interactions in Membranes

B. de Kruyff

Department of Biochemistry, State University of Utrecht, Padualaan 8, De Uithof, Utrecht, The Netherlands

Sterols are main constituents of many cellular membranes. In membranes of mammalian cells, the major sterol present is cholesterol. In erythrocyte and myelin membranes, high levels of cholesterol are found; the molar ratios of cholesterol to phospholipid in these membranes is close to 1.0 (1, 2). Lower amounts (molar ratios 0.1 to 0.6) of cholesterol are found in other plasma membranes and subcellular membranes of nuclei, mitochondria, and microsomes (3, 4). Besides cholesterol, relatively small amounts of a few related sterols are also found in mammalian membranes, e.g., 7-dehydrocholesterol (5), cholestanol (6), lathosterol (7), and desmosterol (8). The main sterols in membranes of plants are stigmasterol, sitosterol, and ergosterol.

In order to understand the function of sterols in biologic membranes, studies in model membrane systems have been undertaken. The interactions of cholesterol and phosphatidylcholine (PC) have been the subject of most of these studies. It was demonstrated in monolayers at the air-water interface that cholesterol reduces the molecular area of various PCs (9–12). Also, in the membrane of the liposome, a reduction in the mean molecular area of egg PC by cholesterol was demonstrated using X-ray techniques (13–15). Using this model membrane system, it was further demonstrated in electron spin resonance (esr) (16, 17) and nuclear magnetic resonance (nmr) studies (18–21) that the fatty acid chains of PCs, which are in the disordered liquid-crystalline state, become less mobile due to the hydrophobic interactions between cholesterol and PC (condensing effect of cholesterol). In contrast the fatty acid chains of PCs that are in the ordered gel state become more mobile in the presence of cholesterol (liquefying effect of cholesterol) such that an intermediate state of fluidity is produced by cholesterol that is independent on the physicochemical state of the membrane lipids (22). As a result the heat content of the gel → liquid-crystalline phase transition in various PCs was found to be decreased by the presence of cholesterol (15, 23–25).

These effects influence the passive diffusion of solutes through the lipid bilayer as demonstrated by studies using liposomes. Cholesterol decreased the permeability of bilayers prepared from PC species with the fatty acid chains in the liquid-crystalline state (26, 27), whereas cholesterol increases the nonelectrolyte permeability of PC liposomes with the fatty acid chains in the gel state (28).

In order to get a better understanding of the lipid-sterol interaction it is necessary to know which structural parameters of both the sterol and the lipid molecule are important for the lipid-sterol interaction. Furthermore, since biologic membranes are composed of various molecular species of different lipid classes, it is important to understand the lipid-sterol interaction in bilayers composed of mixtures of various lipids.

EFFECT OF STEROL STRUCTURE UPON THE LIPID-STEROL INTERACTION

The chemical configuration of the sterol molecule was found to be of critical importance for the lipid-sterol interaction in model membranes. A 3β-OH group, a planar ring system (A and B ring transoriented) and a hydrophobic side chain at C-17 were found to be absolute requirements for lipid-sterol interactions as measured in monolayers (29) and in differential-scanning calorimetry (24), permeability (30), and esr (17) studies on liposomes.

To investigate whether the chemical configuration of the sterol molecule is also important for the interaction with other membrane lipids in a biologic membrane, permeability experiments were performed using *Acholeplasma laidlawii* cells. In the absence of cholesterol in the growth medium, the *A. laidlawii* cell membrane contains no cholesterol (Table 1). When cholesterol is present in the growth medium it is incorporated unaltered in the membrane up to 8% of the total lipids without changing the fatty acid composition of the membrane lipids (Table 1), the lipid-protein ratio or the phospholipid:glycolipid ratio of the membrane (24). The cholesterol incorporation significantly decreases the glycerol and erythritol permeability of the cell membrane. This was measured both by determining the initial swelling rate $(d^1/A/dt)\%$ of the cells when they are transferred from an isotonic sucrose solution to an isotonic glycerol or erythritol solution (Fig. 1) and measuring the [^{14}C]erythritol flux through the membrane (31). Since the mean molecular area of the isolated *A. laidlawii* membrane lipids in a monolayer at the air-water interface is reduced by cholesterol (24), it can be concluded that because of the condensing effect of cholesterol membrane permeability is reduced.

Acholeplasma laidlawii cells grown in media supplemented with various sterols incorporate these sterols in their membranes in a similar manner as cholesterol (24, 31). Table 2 shows the effect of the incorporation of various sterols on the glycerol and erythritol permeability through the membrane. The 3α-OH isomers epicholesterol and epicholestanol and the sterol with a *cis*-structured steroid nucleus coprostanol decrease the membrane permeability much less than the 3β-OH isomers with a flat steroid nucleus, cholesterol, cholestanol, and ergosterol. Stigmasterol does not influence the membrane permeability to a marked extent. Thus, as in the model membrane systems, a reduction in permeability of the cell membrane requires a 3β-OH group and a planar configuration of the sterol molecule. It is important to note that those sterols that showed no or poor interaction with other lipids in the *A. laidlawii* cell membrane also could not support growth of the sterol-requiring mycoplasmatales (32, 33).

TABLE 1. *Effect of cholesterol incorporation upon the fatty acid composition of the total lipids from A. laidlawii*

Cholesterol (mg/liter growth medium)	Fatty acid composition of total lipids[a]							Amount of cholesterol incorporated (wt % of total lipid)
	12:0	14:0	16:0	18:0	18:1$_c$	18:2	others	
0	3.4 ± 0.5	9.9 ± 0.8	48.8 ± 1.9	2.7 ± 1.8	35.8 ± 2.0	2.2 ± 0.6	1.3 ± 0.1	0.0
25	3.6 ± 0.3	8.2 ± 0.7	46.1 ± 1.5	2.3 ± 0.4	37.5 ± 1.9	2.3 ± 0.6	1.6 ± 0.3	8.0 ± 0.9[b]

[a] Cells grown on 0.03 mM palmitic acid (16:0) and 0.03-mM oleic acid (18:1c). Mean of 11 experiments given in mole% (±SD).
[b] Mean of eight experiments.

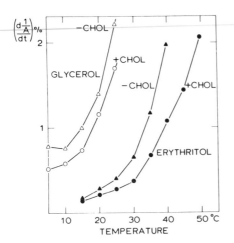

FIG. 1. Initial swelling rates of *A. laidlawii* cells in isotonic glycerol and erythritol. A, Absorbance of the solution at 450 nm. Cells were grown in a medium containing 0.03 mM palmitic and 0.03 mM oleic acid with or without 0.06 mM cholesterol.

EFFECT OF THE POLAR HEADGROUP UPON THE LIPID-CHOLESTEROL INTERACTION

Since the lipid-sterol interaction requires a 3β-OH group of the sterol molecule, interactions between this hydroxyl group and the polar part of the lipid molecule must be considered. Therefore the interactions between cholesterol and a number of synthetic and natural lipids that differ in the polar part of the molecule were studied. The lipid-cholesterol interaction was observed either as a reduction in the mean molecular area in monolayers or as a reduction in the heat content of the gel → liquid-crystalline phase transition in liposomes. The chemical structure of some of the lipids studied are shown in Fig. 2.

The force-area curves of monolayers of 1-palmitoyl-2-oleoyl-*sn*-glycero-3-phosphorylcholine, compound I and compound II, are shown in Fig. 3. The pure lipids, differing from each other by having an ester, ether, or alkane linkage at the 2-position, show similar force-area curves, and are condensed similarly by cholesterol. This result demonstrates that the two oxygen atoms of the ester linkage at the 2-position of the PC molecule are not required for the condensing effect of cholesterol.

The force-area curves of the phosphinate lipids III and IV were also significantly condensed by cholesterol (34), eliminating the possibility that the carbonyl oxygen of the ester-linkage at the 1-position and the oxygen atoms connecting phosphorus to carbon are required for the condensing effect. A natural lipid without the phosphorylcholine group is monoglycosyldiglyceride. When the interaction of this lipid isolated from *A. laidlawii* with cholesterol was measured in the monolayer it was apparent that the phosphorylcholine group is not required for the condensing effect since a marked reduction in mean molecular area was observed (34).

TABLE 2. *Effect of different sterols incorporated in the A. laidlawii cell membrane on the glycerol and erythritol permeability through the membrane*

Sterol incorporated in the membrane	Relative [14C]erythritol equilibrium flux [a]			Relative initial swelling rate in isotonic glycerol			Relative initial swelling rate in isotonic erythritol [a]		
	10°C	15°C	20°C	10°C	20°C	30°C	20°C	30°C	40°C
None	37	60	100	53	100	206	100	195	434
Cholesterol	28	43	66	29	60	—	53	133	324
Cholestanol	33	51	71	43	74	172	63	133	340
Ergosterol	30	47	75	47	80	142	79	165	328
Stigmasterol	37	53	89	49	85	157	88	176	351
Epicholesterol	34	52	—	50	95	—	89	200	450
Epicholestanol	41	60	87	50	93	203	93	183	431
Coprostanol	41	59	90	41	105	201	80	280	384

Cells were grown on 0.03 mM palmitic acid, 0.03 mM oleic acid, and 0.06 mM of the indicated sterols.
[a] All data are related to the permeability values found at 20°C for the sterol free cells.

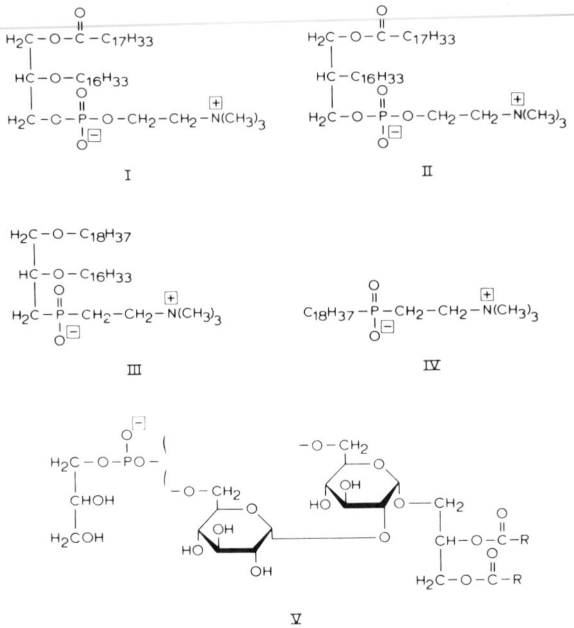

FIG. 2. Chemical structures of various lipids. **(I)** *rac*-1-oleoyl-2-hexadecylglycero-3-phos-phorylcholine; **(II)** *rac*-1-oleoyl-2-*C*-hexa-decylpropanediol-3-phosphorylcholine; **(III)** 2-hex-adecoxy-3-decoxypropyl[2′- (trimethyl-ammonium)ethyl]-phosphinate; **(IV)** octa-decyl 2′(tri-methylammonium)ethyl-phosphi-nate; **(V)** 3-*su*-glycerol-3-phosphoryl-6′-[*O*-α-D-glucopyran-osyl-(1 → 2)-*O*-α-D-gluco-pyranosyl]-*sn*-1,2-diacylglycerol. **I, II, III,** and **IV** are synthetic lipids, **V** was isolated from *A. laidlawii* cell membranes.

The gel → liquid-crystalline phase transition in liposomes prepared from the various lipids shown in Fig. 2 was completely eliminated by the incorporation of 50 mole% of cholesterol (34). Figure 4 shows this effect for lipid V and some other pure lipids isolated from *A. laidlawii*. Thus for both the condensing and liquefying effect of cholesterol none of the oxygen atoms in the ester linkages or the phosphorylcholine group are essential. Thus the requirement of a 3β-OH group in the sterol molecule for interaction with other lipids cannot be explained by any specific interaction in the polar head group region. The solubility of cholesterol and epicholesterol in lipid bilayer, as measured by X-ray (34), is similar. Segregation of epicholesterol in the lipid bilayer in clusters composed of a relatively small number of molecules cannot be excluded. The only difference in force-area curves of monolayers of cholesterol and epicholesterol is a reduced collapse pressure for the latter (29), which might be caused by a different hydra-tion of the 3-OH group. Therefore it is tempting to speculate that the hydration of the 3β-OH-group of sterols is important for the lipid-sterol interaction.

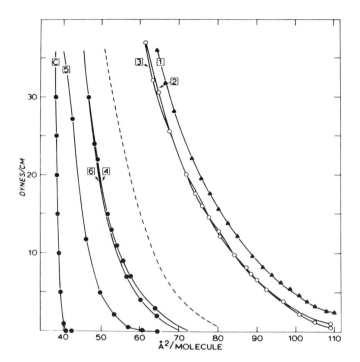

FIG. 3. Force-area curves at 22°C at the air-water interface. **(1)** rac-1-oleoyl-2-hexadecylglycero-3-phosphorylcholine; **(2)** 1-palmitiyl-2-oleoyl-*sn*-glycero-3-phosphorylcholine; **(3)** rac-1-oleoyl-2-C-hexadecyl propane diol-3-phosphorylcholine. **(4–6)** 1:1 mixtures of 1–3 with cholesterol (c). The dotted line represents the calculated curve of assuming that no interaction occurs.

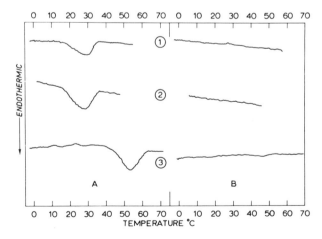

FIG. 4. Thermograms of various pure lipids isolated from *A. laidlawii.* **(1)** Phosphatidylglycerol, **(2)** phosphoglycolipid (formula V), **(3)** diglycosyldiglyceride, in the absence **(A)** or presence **(B)** of 50 M% cholesterol.

LIPID-CHOLESTEROL INTERACTIONS IN MIXTURES OF VARIOUS LIPIDS

Cholesterol can interact with many molecular species of various lipids. However, it was not known whether cholesterol would show a preference for any of the phospholipid species present in membranes composed of various molecular species of lipids. Therefore the effect of cholesterol upon the phase transition(s) occurring in varying mixtures of synthetic PCs was studied (35). The synthetic lipids used were 1,2-dilauryl-*sn*-glycerol-3-phosphorylcholine (12:0/12:0-PC), 1,2-dimyristoyl-*sn*-glycero-3-phosphorylcholine (14:0/14:0-PC), 1,2-dipalmitoyl-*sn*-glycero-3-phosphorylcholine (16:0/16:0-PC), 1,2-distearoyl-*sn*-glycero-3-phosphorylcholine (18:0/18:0-PC), and 1,2-dioleoyl-*sn*-glycero-3-phosphorylcholine (18:1$_c$/18:1$_c$-PC).

Lipid bilayers composed of equimolar amounts of 12:0/12:0-PC—14:0/14:0-PC, 14:0/14:0-PC—16:0/16:0-PC, and 16:0/16:0-PC—18:0/18:0-PC show cocrystallization of the paraffin chains (36–38). This is shown in Fig. 5 for the 16:0/16:0-PC—18:0/18:0-PC mixture. Only one phase transition is observed which occurs at a temperature intermediate between the transition temperatures

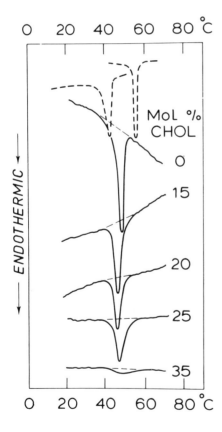

FIG. 5. Thermograms of the equimolar 16:0/16:0-PC—18:0/18:0-PC mixture containing increasing percentages of cholesterol. The dotted curves represent the phase transitions of pure 16:0/16:0-PC and 18:0/18:0-PC.

of the individual PCs. Increasing amounts of cholesterol reduce the energy content of the phase transition of the mixture in a similar manner to that observed for the pure PCs (23, 34, 35). At cholesterol concentrations between 30 and 40 mole% the transition vanished. Apparently, at these concentrations, all PC molecules are interacting with cholesterol. Cholesterol incorporation does not shift the temperature of the phase-transition in the mixture significantly, demonstrating that cholesterol interacts randomly with both molecular species present. Similar effects were observed for the other mixtures that showed cocrystallization of the paraffin chains.

Bilayers composed of mixtures of equimolar amounts of PCs with fatty acid chains differing by at least four carbon atoms or with fatty acid chains differing in degree of unsaturation show monotectic behavior (immixibility in the gel phase). In these mixtures two transitions are present, illustrated in Fig. 6, for the 12:0/12:0-PC—18:0/18:0-PC mixture. Since two transitions are observed in the thermogram, it is possible to unambiguously demonstrate whether cholesterol will show any preference for either of the two molecular species present. As is obvious from Fig. 6, cholesterol at lower concentrations predominantly interacts with 12:0/12:0-PC and only interacts with 18:0/18:0-PC at higher concentrations (above 20 mole%). Other monotectic mixtures exhibit essentially similar behavior. It is concluded therefore that at low concentrations cholesterol interacts preferentially with the lipid species that has the lowest melting temperature. At higher cholesterol concentrations the heat content of the lipid species with the higher melting temperature starts to decrease. The interpretation of these results is that when lateral phase separation occurs, cholesterol tends to associate with the lipids that are in the liquid-crystalline state. This means that at appropriate temperatures in monotetic mixtures cholesterol is nonrandomly distributed. This finding may have direct complications for biologic membranes with low cholesterol content in which phase transitions were detected, such as the mitochondrial

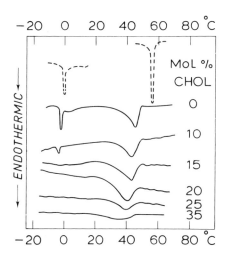

FIG. 6. Thermograms of the equimolar 12:0/12:0-PC—18:0/18:0-PC mixture containing increasing percentages of cholesterol. The dotted curves represent the phase transitions of pure 12:0/12:0-PC and 18:0/-18:0-PC.

(39), microsomal (39), and *A. laidlawii* cell membrane (24). Cholesterol in these membranes might be nonrandomly distributed in the plane of the membrane in situations where lateral phase separation of the membrane lipids occurs.

ACKNOWLEDGMENTS

The stimulating discussions on lipid-sterol interactions in the biomembranes group of the department of Biochemistry in Utrecht, in particular with Dr. R. A. Demel, have been of great help in this study.

REFERENCES

1. Nelson, G. J. (1967): *J. Lipid Res.,* 8:374.
2. Demel, R. A., London, Y., Geurts van Kessel, W. S. M., Vossenberg, F. G. A., and van Deenen, L. L. M. (1973): *Biochim. Biophys. Acta,* 311:507.
3. Asworth, L. A. E., and Green, C. (1966): *Science,* 151:210.
4. Lee, T., Stephens, N., Moehl, A., and Snijder, F. (1973): *Biochim. Biophys. Acta,* 291:86.
5. Glover, J., and Green, C. (1957): *Biochem. J.,* 67:308.
6. Werbin, H., Chaikoff, I. L., and Imada, M. R. (1962): *J. Biol. Chem.,* 237:2072.
7. Miller, W. L., and Baumann, C. A. (1954): *Proc. Soc. Exp. Biol. Med.,* 85:561.
8. Smith, M. E., Fumagalli, R., and Paoletti, R. (1967): *Life Sci.,* 6:1085.
9. de Bernard, L. (1958): *Bull. Soc. Chim. Biol.,* 40:161.
10. Demel, R. A., van Deenen, L. L. M., and Pethica, B. A. (1967): *Biochim. Biophys. Acta,* 135:11.
11. Joos, P., and Demel, R. a. (1969): *Biochim. Biophys. Acta,* 183:447.
12. Demel, R. A., Geurts van Kessel, W. S. M., and van Deenen, L. L. M. (1972): *Biochim. Biophys. Acta,* 266:26.
13. Rand, R. P., and Luzzati, V. (1968): *Biophys. J.,* 8:125.
14. Lecuyer, H., and Dervichan, D. G. (1969): *J. Mol. Biol.,* 45:39.
15. Ladbrooke, B. D., Williams, R. M., and Chapman, D. (1968): *Biochim. Biophys. Acta,* 150:333.
16. Hubbel, W. L., and McConnel, H. M. (1971): *J. Am. Chem. Soc.,* 93:314.
17. Butler, K. W., Smith, I. C. P., and Schneider, H. (1970): *Biochim. Biophys. Acta,* 219:514.
18. Dakke, A., Finer, E. G., Flook, A. G., and Phillips, M. C. (1971): *FEBS Lett.,* 18:326.
19. Oldfield, E., Chapman, D., and Derbyshire, W. (1971): *FEBS Lett.,* 16:102.
20. Darke, A., Finer, E. G., Flook, A. G., and Phillips, M. C. (1972): *J. Mol. Biol.,* 63:265.
21. Metcalfe, J. C., Birdsall, N. J. M., and Lee, A. G. (1972): In: *Proc. 8th FEBS Meeting, Amsterdam* p. 197. North Holland, Amsterdam.
22. Phillips, M. C. (1972): In: *Progress in Surface and Membrane Science,* edited by J. F. Danielli, M. D. Rosenberg, and D. A. Cadenhead, p. 139. Academic Press, New York.
23. Hinz, H. J., and Sturtevant, J. M. (1972): *J. Biol. Chem.,* 247:3697.
24. de Kruyff, B., Demel, R. A., and van Deenen, L. L. M. (1972): *Biochim. Biophys. Acta,* 255:331.
25. Norman, A. W., Demel, R. A., de Kruyff, B., and van Deenen, L. L. M. (1972): *J. Biol. Chem.,* 247:1918.
26. de Gier, J., Mandersloot, J. G., and van Deenen, L. L. M. (1970): *Biochim. Biophys. Acta,* 219:245.
27. Demel, R. A., Kinsky, S. C., Kinsky, C. B., and van Deenen, L. L. M. (1968): *Biochim. Biophys. Acta,* 150:655.
28. de Gier, J., Mandersloot, J. G., and van Deenen, L. L. M. (1969): *Biochim. Biophys. Acta,* 173:143.
29. Demel, R. A., Bruckdorfer, K. R., and van Deenen, L. L. M. (1972): *Biochim. Biophys. Acta,* 255:311.
30. Demel, R. A., Bruckdorfer, K. R., and van Deenen, L. L. M. (1972): *Biochim. Biophys. Acta,* 255:321.
31. de Kruyff, B., de Greef, W. J., van Eyk, R. V. W., Demel, R. A., and van Deenen, L. L. M. (1973): *Biochim. Biophys. Acta,* 298:479.
32. Smith, P. F. (1964): *J. Lipid Res.,* 5:121.
33. Rottem, S., Pfendt, E. A., and Hayflick, L. (1971): *J. Bacteriol.,* 105:323.

34. de Kruyff, B., Demel, R. A., Slotboom, A. J., van Deenen, L. L. M., and Rosenthal, A. F. (1973): *Biochim. Biophys. Acta,* 307:1.
35. de Kruyff, B., van Dijck, P. W. M., Demel, R. A., Schuyff, A., Brants, F., and van Deenen, L. L. M. (1974): *Biochim. Biophys. Acta,* 356:1.
36. Phillips, M. C., Ladbrooke, B. D., and Chapman, D. (1970): *Biochim. Biophys. Acta,* 196:35.
37. Veruegaert, P. H. J. T., Verkleij, A. J., Elbers, P. F., and van Deenen, L. L. M. (1973): *Biochim. Biophys. Acta,* 311:320.
38. Shimshick, E. J., and McConnel, H. M. (1973): *Biochemistry,* 12:2351.
39. Blazy, K. J. F., and Steim, J. M. (1972): *Biochim. Biophys. Acta,* 266:737.

Lipids, Vol. 1, edited by R. Paoletti, G. Porcellati, and G. Jacini. Raven Press, New York © 1976.

Freeze-Etching and Membranes

P. H. J. Th. Ververgaert* and A. J. Verkley†

Biological Ultrastructure Research Unit and †Laboratory of Biochemistry, State University of Utrecht, Transitorium 3, Utrecht, The Netherlands

In cell membranes and liposomes, phase transitions from the liquid-crystalline to the gel state can be demonstrated by various physical techniques (1, 2). The morphologic consequence at a supramolecular level can be investigated by freeze-fracture (etch) electron microscopy, which allows a three-dimensional view in the membrane. Especially strains of *Acholeplasma laidlawii* and *Escherichia coli* K1060 are suited for studies of phase transitions in membranes. By growing these cells in the presence of different fatty acids, the fatty acid pattern of membrane lipids and consequently the transition temperature can be varied.

When the material is quenched from above the transition temperature and freeze-fractured, membrane fracture faces are covered with randomly distributed particles (3, 4). Material that is chilled to a temperature below the transition and thereafter quenched shows strongly aggregated particles. Within the transition the extent of the particle aggregation increases as more lipid is solid (5).

Because intramembranous particles are related to proteins or glycoproteins, it was inferred from the observed particle aggregation that the solidification of the acyl moiety of the membrane results in a lateral squeezing out of proteins and glycoproteins. A large number of microorganisms, all of which do not possess cholesterol as a membrane constituent showed similar phenomena as found in *A. laidlawii* and *E. coli* (4, 6).

However, membranes like those of erythrocytes are not liable to such ultra-structural modifications because of the presence of cholesterol. It has been established by differential scanning calorimetry and nuclear magnetic resonance (nmr) that this compound creates a semiliquid-crystalline state and prevents a crystallization of the phospholipid when artificial or biomembranes are cooled (7–9).

That cholesterol can prevent particle aggregation has been exemplified in a comparative study of a cholesterol-containing native strain of *Mycoplasma mycoides* and a temperature-adapted strain of this mycoplasma with a substantially lower cholesterol concentration (10). Recently it was found that also microorganisms containing branched acyl chains in membranes did not show particle aggregation, although a clear thermotropic transition could be measured (4). The reason for the nonappearance of particle aggregation upon solidification of the lipids was ascribed to the fact that the packing of the acyl chains was not changed during the transition as observed by X-ray diffraction.

To understand in more detail the lipid phase transition in a biologic membrane of mixed lipid composition, we studied the crystallization process in liposomes prepared from a single phosphatidylcholine (PC) species and from mixtures of two species. Freeze-fracture electron microscopy has demonstrated that liposomes prepared from synthetic PCs have smooth or slightly rugged fracture faces above the transition temperature (3, 11, 12), whereas below that temperature characteristic band patterns are observed for each PC species. From X-ray diffraction studies (13), it was concluded that under these circumstances at least two PCs, which were also studied by freeze etching, dilauroyl- and dimyristoyl-lecithin, are organized in undulated bilayers in which the molecules are rigid and tilted with respect to the normal of the bilayer plane. The visualization of this so-called $p\beta'$ configuration as arrays of light and dark bands in freeze-etch replicas (Fig. 1a) enables the detection and examination of phase separation in lecithin mixtures by freeze-etch electron microscopy (14, 15).

In studying the influence of cholesterol on the crystallization of one PC species we found that lecithin-cholesterol mixtures above a concentration of 20 Mole% cholesterol displayed smooth fracture faces when the lipid material was quenched below the transition temperature (Fig. 1d). In the presence of 10 and 15 Mole% of cholesterol, a homogenous band pattern is visible, when liposomes in the gel state were quenched. No smooth areas indicating cholesterol-PC complexes next to areas demonstrating a band pattern of free PC could be found (Fig. 1b,c).

It is known from differential scanning calorimetry that PC mixtures of two species that differ by only two carbon atoms have only one thermotropic peak (1). When cholesterol is added there is no change in the transition temperature but only a decrease in the energy content of the transition, indicating a PC-cholesterol interaction (16).

From these data it was concluded that cholesterol in a cocrystallizing system does not interact specifically with one of the two phospholipids. With freeze etching we found that a mixture of dilauryl- and dimyristoyl-lecithin displayed one regular crystallization pattern when quenched from below the transition. The periodicity (180 Å) significantly differed from those of the individual species. This finding strongly supports the idea that the lipids tend to crystallize simultaneously in one crystalline lattice. When 10 mole% of cholesterol is added to the equimolar mixture a homogenous band pattern is found. No morphologic features suggesting an interaction between cholesterol and only one of the PC species in the mixtures were seen in that. No smooth areas corresponding with cholesterol-PC complexes next to areas with a band pattern characteristic for either of the two phospholipids were present (15).

Also the differential scanning calorimetrical results on mixtures showing a monotectic behavior could be supported by freeze-etch data. Equimolar mixtures of dimyristoyl-lecithin and dioleoyl-lecithin or of dipalmitoyl-lecithin and 1-palmitoyl-2-oleoyl lecithin show two thermotropic peaks in the differential scanning calorimetric curve (16). When for example the last mixture was quenched from +23°C (i.e., between the two peaks) one observes discrete regions with a band pattern corresponding with solid dipalmitoyl-lecithin, suspended in

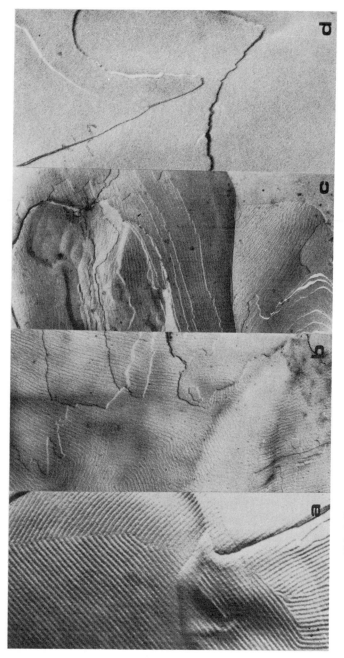

FIG. 1. Fracture faces of liposomes from dimyristoyl PC containing **(a)** 0 mole%, **(b)** 10 mole%, **(c)** 15 mole%, and **(d)** 25 mole% cholesterol quenched from +5°C. *Magnification:* approximately × 60,000.

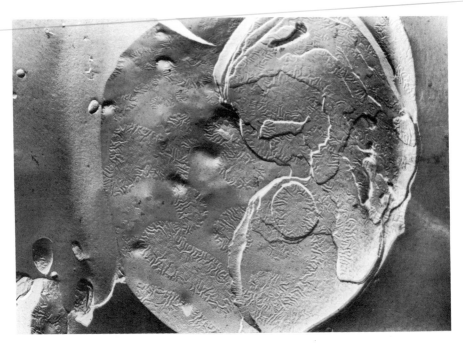

FIG. 2. Fracture faces of an equimolar mixture of dipalmitoyl- and 1-palmitoyl-2-oleoyl-PC quenched from +23°C (i.e., between the two thermotropic peaks). *Magnification:* approximately X 40,000.

the liquid-crystalline 1-palmitoyl-2-oleoyl lecithin (Fig. 2). When the mixture was quenched from −10°C when both species are in the gel state, the individual band patterns of both species are visible.

Addition of 10 mole% of cholesterol to this mixture and quenching from −10°C resulted in fracture faces with band patterns characteristic of dipalmitoyl-lecithin next to smooth areas representing the 1-palmitoyl-2-oleoyl lecithin-cho-lesterol fraction. Hence, in agreement with differential scanning calorimetry it is suggested that cholesterol interacts preferentially with the lowest melting species upon cooling a mixture with monotectic behavior (15, 16).

The usefulness of freeze-etch studies on model membranes has recently been reinforced. Chen and Hubbell succeeded in incorporating rhodopsin in dimyristoyl-lecithin liposomes (17). In their experiments both the distribution of intramembranous particles and the occurrence of the typical dimyristoyl-lecithin band pattern allowed detailed investigation of the lipid-protein interaction.

REFERENCES

1. Ladbrooke, B. D., and Chapman, D. (1969): *Chem. Phys. Lipids,* 3:304.
2. Engelman, D. M. (1971): *J. Mol. Biol.,* 58:153.

3. Verkley, A. J., Ververgaert, P. H. J. T., van Deenen, L. L. M., and Elbers, P. F. (1972): *Biochim. Biophys. Acta,* 288:326.
4. Haest, C. W. M., Verkley, A. J., de Gier, J., Scheek, R., Ververgaert, P. H. J. T., and van Deenen, L. L. M. (1974): *Biochim. Biophys. Acta,* 356:17.
5. James, R., and Branton, D. (1973): *Biochim. Biophys. Acta,* 323:378.
6. Speth, V., and Wunderlich, F. (1973): *Biochim. Biophys. Acta,* 291:621.
7. Ladbrooke, B. D., Williams, R. M., and Chapman, D. (1968): *Biochim. Biophys. Acta,* 150:33.
8. Oldfield, E., Chapman, D., and Derbyshire, W. (1971): *FEBS Lett.,* 16:102.
9. Oldfield, E., Chapman, D., and Derbyshire, W. (1971): *Biochem. Biophys. Res. Commun.,* 43:610.
10. Rottem, S., Yashouv, J., Ne'eman, Z., and Razin, S. (1973): *Biochim. Biophys. Acta,* 323:495.
11. Pinto da Silva, P. (1971): *J. Microsc.,* 12:185.
12. Grant, C. W. M., Wu, S. H. W., and McConnell, H. M. (1974): *Biochim. Biophys. Acta,* 363:151.
13. Tardieu, A., Luzzati, V., and Reman, F. C. (1973): *J. Mol. Biol.,* 75:711.
14. Ververgaert, P. H. J. T., Verkley, A. J., Elbers, P. F., and van Deenen, L. L. M. (1973): *Biochim. Biophys. Acta,* 311:320.
15. Verkley, A. J., Ververgaert, P. H. J. T., de Kruyff, B., and van Deenen, L. L. M. (1974): *Biochim. Biophys. Acta,* 373:495.
16. de Kruyff, B., van Dyck, P. W. M., Demel, R. A., Schuyff, A., Brands, F., and van Deenen, L. L. M. (1974): *Biochim. Biophys. Acta,* 356:1.
17. Chen, Y. S., and Hubbell, W. L. (1973): *Exp. Eye Res.,* 17:517.

Lipids, Vol. 1, edited by R. Paoletti, G. Porcellati,
and G. Jacini. Raven Press, New York © 1976.

The Role of Cholesterol as a Membrane Component: Effects on Lipid-Protein Interactions

Demetrios Papahadjopoulos

Experimental Pathology, Roswell Park Memorial Institute, Buffalo, New York 14203

Cholesterol is a prominent component of mammalian plasma membranes and its role in the biology of the cell is of considerable importance, especially in view of the possible relationship to the etiology of atherosclerosis (1). Studies with well-defined membrane systems (such as monolayers, bilayers and vesicles) have been important in relating the physicochemical properties of phospholipid-cholesterol interactions to the possible physiologic role of cholesterol in cell membranes. Recent reviews (2–4) have discussed in detail the effects of cholesterol on the properties of phospholipid membranes such as the condensation of the area per molecule in monolayers, the inhibition of motion within the outer segment of the phospholipid acyl chains in bilayers, the increase in width of the bilayer and the increased perpendicular orientation of the acyl chains. Finally it has been well documented that the presence of cholesterol generally decreases the permeability of phospholipid bilayers to ions and small polar molecules and the results were reviewed recently (4).

The currently accepted concept for biologic membrane structure (5) involves a lipid bilayer as the basic structural unit, with proteins embedded partly or fully into the interior of the lipid bilayer. A considerable percentage of membrane proteins are also known to be loosely bound to the surface of membranes (6). It is reasonable to expect then that changes in lipid composition might be important in defining the localization and also the conformation of membrane-bound proteins. Following the same argument, it would be reasonable to expect that the presence or absence of cholesterol in membranes could affect not only the properties of the lipid bilayer, as such, but also have considerable effects on the function of membrane proteins.

The approach followed in this laboratory for the last few years (4) has been to study the properties of pure lipid bilayers and how they are affected by different purified proteins. Conversely we have also studied the effects of lipid composition on the function of specific proteins with which they interact. The investigations carried out so far, indicate that the lipid composition, and specifically the presence or absence of cholesterol, can play an important role in defining the localization and function of proteins and enzyme systems in membranes (7, 8). These studies are in good agreement with results obtained by other investigators using intact biologic membranes (9–12). In this chapter the effects of cholesterol on the interaction of several purified membrane proteins with phospholipids are discussed in terms of monolayer expansion, vesicle permeability, and enzymatic activity of a reconstituted adenosinetriphosphatase (ATPase).

MATERIALS AND METHODS

All phospholipids were prepared in this laboratory and were pure by the criterion of thin-layer chromatography (TLC) on silica gel H and a solvent of chloroformmethanol7M ammonia (230:90:15; v/v/v). Phosphatidylserine (PS) from beef brain and phosphatidylcholine (PC) from egg yolk were prepared as described before (13). Dipalmitoylphosphatidylglycerol (DPPG) was synthesized from dipalmitoylphosphatidylcholine (DPPC) by minor modification (14) of the method of Dawson (15). DPPC was synthesized according to Robles and Van den Berg (16). Dioleylphosphatidylglycerol (DOPG) was synthesized similarly to DPPG. Cholesterol was obtained from Sigma and recrystallized twice from methanol. Lipids were stored in chloroform under nitrogen in sealed ampoules at $-50°C$.

Cytochrome c (horse heart, type VI) was obtained from Sigma, albumin (human serum, crystallized) from Mann Research. Gramicidin A (activity 100%) was supplied by Nutritional Biochemicals. Myelin basic protein (A1) was a gift from Dr. E. H. Eylar, and was prepared from bovine brain (17). Myelin proteolipid apoprotein (N-2 protein fraction) was a gift from Dr. M. Moscarello and was prepared from human brain myelin (18). The N-2 was used as a solution in water (19). ATPase was prepared from frozen rabbit kidney outer medulla (obtained from Pell-Freeze Biologicals) as described previously (20).

Methods for phospholipid protein and ATPase determinations have been reported (20). Preparation of sonicated phospholipid vesicles and measurements of Na^+ permeability in the absence and presence of cholesterol and proteins have been described (7, 21, 22). Expansion of monolayers was carried out as described by Papahadjopoulos et al. (7). Chemicals and other details, are also as described in ref. 7.

RESULTS AND DISCUSSION

Permeability of Phospholipid Vesicles: Effects of Cholesterol and Acyl Chain Fluidity

The self-diffusion rates of ions through phospholipid vesicles are extremely low and are further reduced when cholesterol is incorporated into the bilayers (21). The large differences in self-diffusion rates between anions and cations (21) and between K^+ and Na^+ (23) indicate that the vesicles are "tight" and diffusion from the interior to the exterior aqueous spaces does not involve breaking of the membrane, but specific sites. Temperature studies on the effect of cholesterol on vesicle permeability indicated that the activation energy for the diffusion of cations is drastically decreased, whereas that for the anions is slightly increased in the presence of cholesterol (21). The decrease of the activation energy for the diffusion of cations was explained as resulting from a stabilization of the phospholipid bilayer by cholesterol against temperature-dependent structural changes (21). In addition the presence of cholesterol also produced a change from a discontinuous to a linear Arrhenius plot for the diffusion of cations through PS vesicles between 12 and 50°C.

The effects mentioned above were obtained when cholesterol was mixed with phospholipids at temperatures above the transition point (T_c), and the acyl chains were in a fluid state. The stabilizing effect of cholesterol can be shown also at a temperature range that includes the T_c of the pure phospholipid. Earlier studies with a differential scanning calorimeter had produced evidence that the presence of cholesterol at 1:1 molar ratio completely abolishes the endothermic transition of DPPC (2, 3). In this laboratory we have studied the same effect by following the efflux of Na^+ through dipalmitoylphospholipid vesicles (14). As shown in Fig. 1, the self-diffusion rate of $^{22}Na^+$ through DPPG vesicles increases sharply at the temperature range 30 to 40°C. This is the temperature range for the melting of the acyl chains and transition to a liquid-crystalline state for sonicated vesicles of this phospholipid as judged by calorimetry and fluorescence polarization (23). The local maximum of Na^+ diffusion rate near the midpoint of the phase transition was interpreted as resulting from discontinuities between liquid and solid domains within the membranes of vesicles undergoing phase transition (14).

The presence of cholesterol in the same vesicles changes completely the above temperature dependence of the Na^+ diffusion rate. As also shown in Fig. 1 the

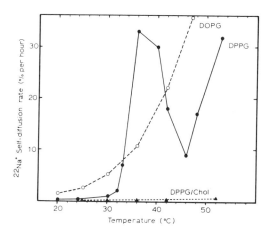

FIG. 1. Self-diffusion rates of $^{22}Na^+$ through phospholipid vesicles at different temperatures (14). ●—●, DPPG; △----△, equimolar mixture of DPPG and cholesterol; ○----○, DOPG. (From Papahadjopoulos et al., ref. 14.)

cholesterol-containing vesicles show a gradual increase in permeability with temperature throughout the phase transition region. The actual diffusion rates are considerably lower at all temperatures than those obtained with the pure lipid without cholesterol and give a linear Arrhenius plot with a slope indicating a 13 kCal/mole activation energy. Vesicles that are fluid throughout this temperature range (DOPG) are also shown in Fig. 1. They show a monotonic increase in diffusion rates with temperature, which gives a linear Arrhenius plot with an activation energy of 24 kCal/mole (14). Addition of cholesterol to this lipid results in vesicles with decreased diffusion rates and a lower activation energy (21).

It is evident from the above results that addition of cholesterol decreases the permeability of phospholipid vesicles to cations and also decreases the Arrhenius activation energy for diffusion. This effect is similar to that obtained by freezing the acyl chains below the phase transition temperature. However, with cholesterol present at 1:1 molar ratio, the vesicles do not undergo the regular phase transition, which is normally accompanied by a sharp increase in permeability.

Penetration of Proteins into Phospholipid Bilayers and Monolayers

Several soluble proteins have been shown to increase the permeability of phospholipid vesicles to Na^+ (22, 24). These effects were later correlated with the ability of such proteins to increase the surface pressure or expand monolayers of the same phospholipids (24, 25). The increase in permeability was attributed to distortion of the bilayer interior, following the establishment of hydrophobic contacts between the protein and the acyl chains. Models for protein localization involving both deformation and penetration of the bilayer have been proposed (25–27) and the subject has been reviewed recently (4).

Further studies on lipid-protein interactions with a differential scanning calorimeter have established three types of interactions (28) that are in good agreement with the previous studies involving vesicle permeability and monolayer expansion. These are summarized in Fig. 2, where the results of all three techniques are compared. Interaction type 1, exemplified by ribonuclease, is depicted as a simple electrostatic adsorption at the phospholipid-water interface. Although it binds strongly, this protein has a minimal effect on vesicle permeability (P), monolayer expansion (A), phase transition temperature (T_c), and enthalpy (ΔH). Cytochrome c and the basic myelin protein (A1) exemplify interaction type 2, which is depicted as partial penetration and deformation of the bilayer. Characteristically these proteins produce a large increase in P and A and a large decrease in T_c and ΔH. Finally interaction type 3, exemplified by the myelin proteolipid apoprotein, is depicted as embedding or deep penetration of a large part of the protein molecule into the bilayer. Such interaction does not depend on electrostatic charges and is characterized by a large increase in P and A, no effect on the T_c, and a decrease in ΔH. Its incorporation into bilayers can be detected by the appearance of intramembranous particles in freeze-fractures (29).

group	protein	lipid	T_c	ΔH	P	A
1	RNase Polylysine	DPPG	= ↑	= ↑	=	=
2	Cyt. c A1	DPPG	↓	↓	↑	↑
3	N-2 Gramicidin	DPPG or DPPC	=	↓	↑	↑

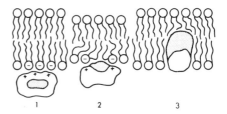

FIG. 2.(A) Summary of the effects of different proteins on properties of phospholipid bilayers and monolayers. Group number indicates the type of interaction shown in **B**. Proteins added to the system are pancreatic ribonuclease (RNAse), cytochrome c (Cyt. c), the basic myelin protein (A1), the major myelin proteolipid apoprotein (N-2), and Gramicidin A (Gramicidin). The transition temperature **(T_c)** from solid to liquid-crystalline form as determined by differential scanning calorimetry (23), the enthalpy (ΔH) of the endothermic transition (23), the permeability (P) of the vesicles (14), and the area per molecule (A) in monomolecular films at the air-water interface (7) are also given. **(B)** Schematic representation of three types of lipid-protein interactions: (1) simple adsorption of the protein at the phospholipid-water interface via ionic bonds; (2) partial penetration and deformation of the bilayer via both nonpolar interactions and ionic bonds; (3) complete penetration of part of the protein molecule into the bilayer at either half length as shown, or spanning the length of the bilayer, via nonpolar interactions. The numbering system coincides with the types of interactions discussed in **A**.

The effect of cholesterol on the interaction of different proteins depends on the type of interaction. As shown in Table 1, incorporation of equimolar amounts of cholesterol produces a large (100-fold) inhibition of the ability of cytochrome c and A1 to increase the Na^+ efflux rate through PS vesicles. On the other hand, it has a minimal effect on the ability of the proteolipid apoprotein to increase the permeability. These results were supplemented with studies on monolayer expan-

TABLE 1. *Effects of various proteins on Na^+ efflux through PS vesicles with or without cholesterol[a]*

Protein	$^{22}Na^+$ self-diffusion rate (% per hr)		Ratio (Column 1/ Column 2)
	PS	PS/Chol.	
None	0.06	0.02	3
Cytochrome c	46.9	0.41	115
Basic myelin protein (A1)	40.6	0.92	44
Major apoprotein of myelin proteolipid (N-2)	15.0	4.9	3

[a] Each reaction mixture contained approximately 1 μmole of PS (with or without 1 μmole of cholesterol) in 1.0 ml of total volume of buffer as sonicated vesicles loaded with $^{22}Na^+$. Proteins were added outside the vesicles at concentrations and conditions indicated below. The experiments have been described in more detail earlier (7). Cytochrome c: 10 mg/ml, in 10-mM NaCl, pH 7.4, 36°C. Basic myelin protein: 0.65 mg/ml, in 100-mM NaCl, pH 7.4, 24°C. Major apoprotein of myelin proteolipid: 0.36 mg/ml, 10-mM NaCl, pH 6.5, 26°C. None: 10- or 100-mM NaCl, pH 7.4, 36°C.

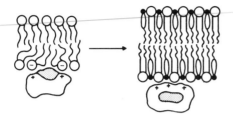

FIG. 3. Schematic representation of a possible freezing-out effect of cholesterol. **(Left)** Phospholipid; **(right)** phospholipid + cholesterol.

sion, which showed that the presence of cholesterol inhibited the expansion by cytochrome c and A1, but not by the proteolipid apoprotein (7).

It appears from the above results that cholesterol inhibits type 2 interactions, perhaps by the stabilization of bilayer discussed earlier. If this is true, simply freezing of the acyl chains by decreasing the temperature below the T_c should have a similar inhibitory effect on the ability of cytochrome c to increase the permeability of vesicles. This was shown to be the case in experiments reported recently, where cytochrome c was added at different temperatures to DPPG and DOPG vesicles with and without cholesterol (7). The effect of cholesterol on type 2 interaction can then be characterized as a freezing-out effect, shown graphically in Fig. 3.

Effects of Cholesterol and Acyl Chain Fluidity on the Activity of (Na+K)-ATPase

The results with the proteolipid apoprotein discussed above, indicate that cholesterol does not inhibit a type 3 interaction. In other words proteins that tend to interact with the bilayer primarily via nonpolar associations (hydrophobic) are not inhibited from penetrating into the bilayer, and perhaps are not excluded from the bilayer interior when the acyl chains are frozen. However, it is still reasonable to expect that changes in the viscosity of the lipid bilayer (induced either by incorporation of cholesterol or by thermotropic transitions) might have pronounced effects on the conformation and function of such membrane-embedded proteins (30). The system we have employed in order to investigate this possibility is a reconstituted (Na + K)-ATPase (8).

A partly purified, delipidized preparation of this enzyme prepared from rabbit kidneys shows very little activity unless PS or PG vesicles are added to it (20). We have used a series of synthetic PG preparations with well-defined melting points and studied the enzyme at different temperatures. The melting points of the sonicated vesicle preparations of DOPG, dimyristoylphosphatidylglycerol (DMPG), DPPG, and distearoylphosphatidylglycerol (DSPG) were investigated by differential scanning calorimetry. Sonication of these phospholipid preparations in dilute suspension produces a lowering of the T_c by two to three degrees and considerable broadening of the endothermic peak (23). The arrows in Fig. 4 indicate the temperature for the initial rise of the endothermic peak, or, the onset of the transition to the fluid state for the pure sonicated vesicles.

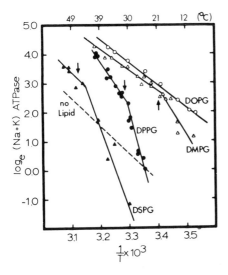

FIG. 4. Arrhenius plots of (Na+K)-ATPase reactivated by the addition of phosphatidyl-glycerol vesicles with different acyl chains. The arrows represent the temperatures of the onset of the endothermic phase transition (melting of the acyl chains) as determined by differential scanning calorimetry. ---, No lipid; ▲, DSPG; ●, DPPG; △, DMPG; ○, DOPG. (From Kimelberg and Papahadjopoulos, ref. 8.)

The enzyme activity obtained with each vesicle preparation at temperatures between 10 and 50°C is shown in Fig. 4 in the form of Arrhenius plots. It can be seen clearly that the fluid vesicles (DOPG) give a straight line, with a slope giving an activation energy of 15 kCal/mole. The other vesicles give definite deviations from linearity with points of discontinuity at 20 (DMPG), 31 (DPPG) and 44°C (DSPG). The temperatures for the discontinuities are in very good agreement with the respective onsets for the transition to the fluid state for each vesicle. At temperatures below this point, the enzyme activity is diminished

TABLE 2. *Effects of cholesterol on the activity of phospholipid-dependent (Na+ + K+)-ATPase*

Phospholipid added[a]	(Na+K)-ATPase (μmoles P$_i$/mg protein · hr)	Percent inhibition
None	3.3	—
PS	37.5	Control
PS/cholesterol (1:1.1)	14.3	62
DOPG	35.2	Control
DOPG/cholesterol (1:1.5)	14.9	58
DPPG	17.8	Control
DPPG/cholesterol (1:1)	1.4	92
PS/PC (1/1)	43.8	—

[a] Numbers in parentheses are molar ratios of the lipids used to make the sonicated vesicles. The amounts added were PS: 0.64 μmoles, DOPG: 0.1 μmoles, DPPG: 0.59 μmoles. Protein 0.07 mg in 1.5-ml total volume. Other details have been described previously (7).

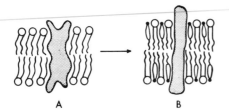

FIG. 5. Schematic representation of a possible freezing-in effect of cholesterol. **(A)** Phospholipid bilayer with penetrating protein. **(B)** phospholipid-cholesterol bilayer, with penetrating protein.

A   B

drastically, indicating that this enzyme is very sensitive to the freezing of the acyl chains of the phospholipid vesicles.

By extrapolation from the above results, it would be expected that incorporation of cholesterol into the phospholipid bilayers would have an inhibitory effect on enzyme activity since it inhibits the motional freedom of the phospholipid acyl chains. The results of such experiments are given in Table 2; cholesterol was found to inhibit DOPG by 58%, PS by 62%, and DPPG by 92%. Simple dilution of the surface charge density of the vesicles by egg PC was not found to be inhibitory.

It thus appears that freezing of the acyl chains, as well as cholesterol incorporation can have substantial inhibitory effects on the activity of this enzyme. Transport enzymes would be expected to be able to span the membrane and there is evidence for an asymmetric localization of the (Na + K)-ATPase on the plasma membrane with different binding sites on either side. Such localization would be characterized essentially similar to interaction type 3 discussed earlier with complete penetration through the lipid bilayer. The dependency on membrane viscosity could be interpreted as a loss of conformational or rotational freedom following freezing of the acyl chain of the bilayer in which the protein is embedded. Such a freezing-in arrangement is depicted graphically in Fig. 5 by the incorporation of cholesterol into the lipid bilayer.

The dependence of membrane enzyme activity on the viscosity of the lipid bilayer was termed viscotropy (8, 20). Several transport and other membrane enzyme systems have been shown to exhibit this dependency (30–33), and it is possible that the phenomenon might be of considerable biologic importance. In addition to the chain length and unsaturation of the acyl chains and cholesterol, membrane fluidity could also be controlled by the interaction of divalent metals with acidic phospholipids, which tend to stabilize the bilayers and drastically increase the melting points (22, 23, 34, 35).

CONCLUSIONS

The studies reported here have indicated that the presence of cholesterol can inhibit the interaction of many proteins with phospholipid membranes. It would appear that these inhibitory effects of cholesterol are the result of the decreased molecular motion of the acyl chains, which makes the deformation or penetration of the bilayer by proteins energetically unfavorable.

The fluidity of the acyl chains of phospholipids in biologic membranes provides the required motional freedom allowing membrane proteins to undergo conformational, rotational, and/or translational movements associated with their activity. Cholesterol could thus have a regulatory role in these processes by controlling the fluidity of the acyl chains.

An increased or decreased level of cholesterol in membranes due to a pathologic situation could thus interfere with vital functions of the membrane with detrimental effects on cell metabolism and viability. It has been proposed that changes in membrane fluidity could be involved in the regulation of growth and differentiation in normal and malignant cells (36) and also as a triggering factor in the development of atherosclerosis (1).

SUMMARY

The possible role of cholesterol in membranes has been studied by following its effect on lipid-protein interactions in reconstituted systems. It has been found that the presence of cholesterol inhibits proteins such as cytochrome c and the basic myelin protein in their ability to increase the permeability of phospholipid vesicles to Na$^+$, and also to expand the area per molecule of phospholipid monolayers at the air-water interface. Using the same model systems and the same criteria, it has been found also that cholesterol does not inhibit the interaction of the myelin proteolipid apoprotein with phospholipids. Finally it was found that cholesterol can inhibit the activity of a reconstituted (Na + K)-ATPase. Similar inhibition was also obtained with this enzyme when the acyl chains of the phospholipid vesicles used for reactivation were frozen.

It was concluded that cholesterol affects the interaction of proteins with phospholipid bilayers by inhibiting the motion of the phospholipid acyl chains. As a consequence, proteins can be either frozen-out or frozen-in depending on the type of interaction.

REFERENCES

1. Papahadjopoulos, D. (1974): *J. Theor. Biol.,* 43:329.
2. Oldfield, E., and Chapman, D. (1972): *FEBS Lett.,* 23:285.
3. Phillips, M. C. (1972): In: *Progress in Surface and Membrane Science,* edited by J. F. Danielli, M. D. Rosenberg, and D. A. Cadenhead, p. 139. Academic Press, New York.
4. Papahadjopoulos, D. and Kimelberg, H. (1973): In: *Progress in Surface Science, Vol. 4, Part 4,* edited by S. G. Davison, pp. 141–232. Pergamon Press, Oxford.
5. Singer, S. J., and Nicolson, G. (1972): *Science,* 175:720.
6. Singer, S. J. (1974): *Ann. Rev. Biochem.,* 43:805.
7. Papahadjopoulos, D., Cowden, M., and Kimelberg, H. (1973): *Biochim. Biophys. Acta,* 330:8.
8. Kimelberg, H., and Papahadjopoulos, D. (1974): *J. Biol. Chem.,* 249:1071.
9. Bruckdorfer, K. R., Demel, R. A., DeGier, J., and van Deenen, L. L. M. (1969): *Biochim. Biophys. Acta,* 183:334.
10. Kroes, J., and Ostwald, R. (1971): *Biochim. Biophys. Acta,* 249:647.
11. DeKruyff, D., Demel, R. A., and van Deenen, L. L. M. (1972): *Biochim. Biophys. Acta,* 255:331.
12. Rottem, V., Cirillo, P., DeKruyff, B., Shinitzky, M., and Razin, S. (1973): *Biochim. Biophys. Acta,* 323:509.

13. Papahadjopoulos, D., and Miller, N. (1967): *Biochim. Biophys. Acta,* 135:330.
14. Papahadjopoulos, D., Jacobson, K., Nir, S., and Isac, T. (1973): *Biochim. Biophys. Acta,* 311:330.
15. Dawson, R. M. C. (1967): *Biochem. J.,* 102:205.
16. Robles, E. C., and van den Berg, D. (1969): *Biochim. Biophys. Acta,* 187:520.
17. Oshiro, H., and Eylar, E. H. (1970): *Arch. Biochem. Biophys.,* 138:392.
18. Gagnon, J. H., Finch, P. R., Wood, D. D., and Moscarello, M. A. (1971): *Biochemistry,* 10:4756.
19. Anthony, J., and Moscarello, M. A. (1971): *FEBS Lett.,* 15:335.
20. Kimelberg, H. K., and Papahadjopoulos, D. (1972): *Biochim. Biophys. Acta,* 282:277.
21. Papahadjopoulos, D., Nir, S., and Ohki, S. (1972): *Biochim. Biophys. Acta,* 266:561.
22. Kimelberg, H. K., and Papahadjopoulos, D. (1971): *J. Biol. Chem.,* 246:1142.
23. Jacobson, K., and Papahadjopoulos, D. (1975): *Biochemistry,* 14:152.
24. Juliano, R. L., Kimelberg, H. K., and Papahadjopoulos, D. (1971): *Biochim. Biophys. Acta,* 241:894.
25. Kimelberg, H. K., and Papahadjopoulos, D. (1971): *Biochim. Biophys. Acta,* 233:805.
26. Gulik-Krzywicki, T., Shechter, E., Luzzati, V., and Faure, M. (1969): *Nature,* 223:1116.
27. Rand, R. P. (1971): *Biochim. Biophys. Acta,* 241:823.
28. Papahadjopoulos, D., and Isac, T. (1974): *Fed. Proc.,* 33:1254, Abst. 167.
29. Vail, W. J., Papahadjopoulos, D., and Moscarello, M. (1974): *Biochim. Biophys. Acta,* 345:463.
30. Raison, J. K. (1973): *Bioenergetics,* 4:285.
31. Wilson, G., and Fox, C. F. (1971): *J. Mol. Biol.,* 55:49.
32. Overath, P., Schairer, H. V., and Stoffel, W. (1970): *Proc. Natl. Acad. Sci. USA,* 67:606.
33. Esfahani, M., Limbrick, A. R., Knutton, S., Oka, T., and Wakil, S. J. (1971): *Proc. Natl. Acad. Sci. USA,* 68:3180.
34. Träuble, H., and Eibl, H. (1974): *Proc. Natl. Acad. Sci. USA,* 71:214.
35. Verkleij, A. J., DeKruyff, B., Ververgaert, P. H. J. T., Tocanne, J. F., and van Deenen, L. L. M. (1974): *Biochim. Biophys. Acta,* 339:432.
36. Vlodavsky, I., and Sachs, L. (1974): *Nature,* 250:67.

Lipids, Vol. 1, edited by R. Paoletti, G. Porcellati, and G. Jacini. Raven Press, New York © 1976.

The Use of Fatty Acid Monolayers as Models for Biomembranes: Autoxidation Studies

James F. Mead and Guey-Shuang Wu

Laboratory of Nuclear Medicine and Radiation Biology, University of California and Department of Biological Chemistry, University of California School of Medicine, Los Angeles, California 90024

The study of the properties of biomembranes is aided by the use of model systems that permit certain of the complex reactions of the membranes to be investigated in a simpler arrangement. We were led to the use of fatty acid monolayers as models by the observation that unsaturated fatty acids, when adsorbed on silica surfaces at a low ratio of fatty acid to adsorbent, are protected against autoxidation under conditions that would result in complete destruction of the unsaturated centers of the fatty acids in some other arrangement (1, 2). As a matter of fact, investigations of fatty acid autoxidation by Porter et al. (3) and in this laboratory have shown that the rate of autoxidation of unsaturated fatty acids is maximal in the monolayer arrangement, falling off as the ratio of fatty acid to adsorbent increases or decreases markedly from that of the complete monolayer.

In Fig. 1 it can be seen that the rate of autoxidation remains constant throughout a wide range of ratios of linoleate to silica, decreasing sharply below 20 mg/g. The meaning of the shape of this curve will become apparent with further discussion.

Although the autoxidation reaction for unsaturated fatty acids is now very well

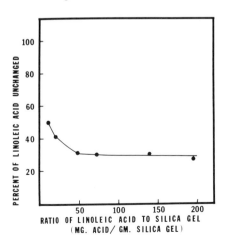

FIG. 1. Extent of autoxidation of linoleic acid adsorbed on silica gel in different ratios of linoleic acid to silica. Autoxidation was carried out for 3 hr at 60°C in air. Ratio for a complete monolayer 190 mg/g.

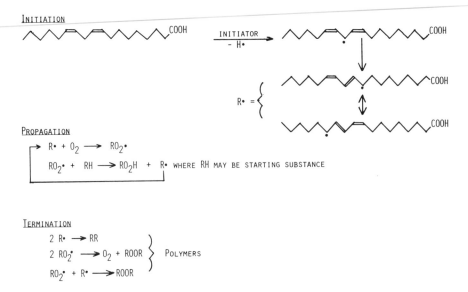

FIG. 2. Simplified scheme of autoxidation of linoleic acid.

understood, certain aspects of the present discussion warrant a brief review of
its mechanism. In Fig. 2 it can be seen that the propagation reaction, by a radical
chain process involving molecular oxygen, is responsible for by far the greatest
yield of product. The initiation reaction, on the other hand, is slow in the absence
of preformed peroxide or catalysis by metal ions or certain other catalysts and
the products of this reaction will, in general, not be apparent in the overwhelming
bulk of products of the propagation steps. As a matter of fact, it can readily be
seen that if unsaturated fatty acids are held in an arrangement in which the
unsaturated centers are kept in close proximity, the chain reaction should be
augmented. The typical membrane phosphoglyceride, usually containing one
saturated and one unsaturated fatty acid, can be constrained into such an arrangement in mono- or bilayer. In Fig. 3A a hypothetical all-saturated orderly arrange-

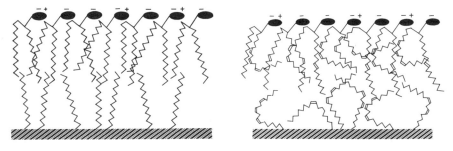

FIG. 3. Schematic arrangement of fatty acids in a phospholipid monolayer or half-bilayer in two
hypothetical cases: **(a)** all saturated fatty acids; **(b)** all polyunsaturated fatty acids.

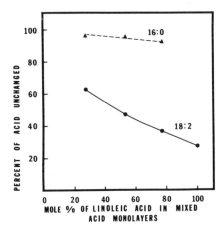

FIG. 4. Extent of autoxidation of mixtures of linoleic and palmitic acids adsorbed on silica gel. Autoxidation was carried out for 3 hr at 60°C in air.

ment is shown, whereas in Fig. 3B a possible representation of an all-polyunsaturated arrangement is depicted. Undoubtedly the true state in most membrane bilayers is somewhere between.

We can approximate this situation in the fatty acid monolayer by carrying out the autoxidation with mixtures of saturated and unsaturated acids. For example, in Fig. 4 the results of such an experiment with mixtures of linoleic and palmitic acids are shown. Comparison with Fig. 1 shows that the palmitic acid has a marked protective effect on the linoleic acid and that this is exerted throughout the range of concentrations. Evidently the mere inclusion of a relatively inert (see curve for 16:0 in Fig. 4) constituent in the monolayer reduces the rate of chain propagation. The 50% ratio might approximate a typical membrane bilayer, in which some protection is afforded simply by the alternate arrangement of saturated and unsaturated chains. Recalling Fig. 1 it can be seen that simple dilution does not afford such protection.

With reference to Fig. 5, it can be seen that in a complete monolayer the unsaturated centers of the fatty acids are in close proximity to each other, facilitating chain propagation. As the ratio of fatty acid to adsorbent decreases, the distance between molecules may increase to the point at which no molecules are contiguous or patches of packed monolayer may remain until an extremely low ratio is reached. It can be seen in Fig. 1 that the latter situation exists and that dilution affords no protection until a ratio of about 25 to 30 mg/g is reached.

FIG. 5. Schematic representation of a linoleic acid monolayer with one fatty acid molecule separated by too great a distance to interact rapidly.

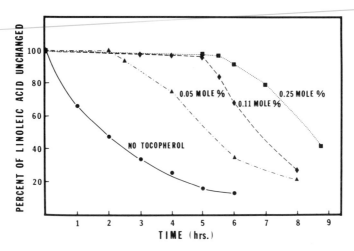

FIG. 6. Effect of different proportions of α-tocopherol on the autoxidation of linoleic acid monolayers.

Motility of the fatty acids in the monolayer probably enters into the picture but the extent of its influence is not yet clear.

Turning now to another facet of membrane protection, we have investigated the effect of α-tocopherol on the rate of autoxidation of linoleic acid monolayers. This study is particularly pertinent since it is now thought that the major function of tocopherol is in the protection of the membrane lipids from autoxidation (4). Indeed Porter et al. (3) have shown that protection against autoxidation is greatest in lipid monolayers (or other ordered arrangement), leading us to question whether the testing of the antioxidant properties of the tocopherols and other physiologic antioxidants in bulk phase lipid has much value. In Fig. 6 the effect of different proportions of α-tocopherol on the autoxidation of linoleic acid monolayers is shown. As has been known for a long time, tocopherol in these ratios does not decrease the rate of autoxidation, but introduces an induction period during which the chain-propagation reaction is prevented by reduction of the chain-initiating substances produced in it. It is of interest that the induction period is proportional to the mole% of tocopherol up to a maximally protective ratio somewhat greater than 0.1 mole%, which is about twice the ratio recommended for dietary tocopherol and polyunsaturated fatty acids. Experiments of this sort with cellular membranes should be carried out and may lead to a reassessment of the recommended ratio. Here again, the effect of motility of the fatty acids and tocopherol has not been taken into account, but it can be estimated that at a molar ratio of 0.10, tocopherol molecules may be separated by about 32 fatty acid molecules supposing even distribution.

CH₃(CH₂)₄⬡(CH₂)₉COOH   **FIG. 7.** Structure of fatty acid hydroquinone.

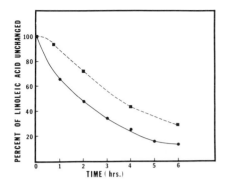

FIG. 8. Effect of 0.1 mole% of fatty acid hydroquinone on the autoxidation of linoleic acid monolayers.

Finally we have made an attempt to tailor an antioxidant that might give maximal protection to fatty acids in monolayer. This compound, shown in Fig. 7, contains a hydroquinone moiety near the center of a fatty acid chain, presumably in the best position to interrupt the radical chain propagation process.[1] In Fig. 8, it can be seen that the results were disappointing since even at the optimal ratio for tocopherol the fatty acid hydroquinone gave almost no induction period. The reasons for this may lie in the binding of both carboxyl and hydroxyls to the adsorbent, placing the antioxidant in an unfavorable position or rendering it immobile. In bulk phase it was about as effective as α-tocopherol. An antioxidant with one of the hydroxyls blocked may avoid this difficulty and is being synthesized.

Thus the use of model systems for the study of biomembranes has already proved to be of considerable interest and it can be predicted that useful information will continue to be gained in the very simple monolayer system.

ACKNOWLEDGMENTS

These studies were supported in part by Atomic Energy Commission Contract AT(04–1)GEN-12 with the University of California and by U.S. Public Health Service Research Career Award GM-K-6-19, 177 from the National Institute of General Medical Sciences.

REFERENCES

1. Slawson, V., and Mead, J. F. (1972): *J. Lipid Res.,* 13:143.
2. Slawson, V., Adamson, A. W., and Mead, J. F. (1973): *Lipids,* 8:129.
3. Porter, W. L., Levasseur, L. A., and Hennick, A. S. (1972): *Lipids,* 7:699.
4. Molenaar, I., Vos, J., and Hommes, F. A. (1972): *Vitamins Horm.,* 30:45.

---

[1] Synthesized by Dr. R. A. Stein of this laboratory.

Lipids, Vol. 1, edited by R. Paoletti, G. Porcellati,
and G. Jacini. Raven Press, New York © 1976.

Phospholipid Composition and Turnover in Neuronal and Glial Membranes

Giuseppe Porcellati and Gianfrancesco Goracci

Istituto di Chimica Biologica, Università di Perugia, Facoltà Medica, C.P. n.3, 06100 Perugia, Italy

The nervous tissue is a very complex system of biologic membranes. Many of them display a number of specialized functions that are peculiar for the nervous system. Since various functional properties of these membranes are known to be related to their physicochemical characteristics, any information related to their composition and turnover appears to be helpful for understanding the relationship between structure and function. It is therefore highly interesting for the biochemist to obtain as much information as possible with regard to the chemical composition and turnover of membrane types in the nervous system.

In recent years several methods have been developed (1–6) for the large-scale preparation of glial and neuronal cells and of their subfractions. These procedures have given the possibility of carrying out *in vitro* several studies on the enzymic activities involved in the synthesis and degradation of membrane-bound phospholipids. It was thus found that, on a protein basis, the neuronal membranes possess a higher base-exchange activity (6), a more relevant phosphorylcholine (PC) and phosphorylethanolamine (PE) diglyceride transferase (7) and a more efficient phospholipase A activity (8) than the glial membranes.

In order to better understand the relationship between the two cell populations in connection with their function, this chapter provides some information on their phospholipid composition and turnover *in vivo* for the more abundant phospholipid classes, choline phosphoglycerides (CPG) and ethanolamine phosphoglycerides (EPG). The turnover of CPG and EPG was followed by injecting rabbits intraventricularly with the correspondent labeled base. At different time intervals after injection, the incorporation of the base into the water-soluble and lipid components was determined in both neuronal and glial cells.

MATERIALS AND METHODS

Surgical Operation

White rabbits (1.5 to 1.8 kg), of both sexes, were anesthetized with sodium pentobarbitone. The calvarium was exposed, and a small hole was made with a dental drill at a point correspondent to the following coordinates: 0.7 cm from choroideal and 0.4 cm from parietal sutures. The hole was then filled with bone wax and the scalp incision closed. The animals were allowed to recover for 4 to 5 days before the isotope injection.

Isotope injection

Local anesthesia was performed by carbocaine injection. The wound was then reopened and a convenient amount of isotope was injected through the hole in one ventricle by means of a MS-100 Terumo microsyringe with a No. 50 B needle. Eighty microliters of 0.9% saline solution containing 9.7 μCi of [1,2-^{14}C]ethanolamine hydrochloride (1.98 mCi/mmole SA) or 60 μl containing 62.35 μCi of [methyl-^{3}H]choline chloride (70.7 mCi/mmole SA) were injected for each experiment.

Neuronal and Glial Cell Fractions

The neuronal and glial fractions were prepared and checked as reported previously (6, 7). The cell-enriched fractions from each brain were homogenized by the use of glass-Teflon homogenizers in 1.5 to 3.0 ml of distilled water at 1,000 revolutions/min with five or six up-and-down strokes.

Extraction of Lipids and Water-Soluble Components

Chloroform-methanol (2:1, v/v) solution (4 ml) were added to the homogenized fractions. After brief agitation, the phases were separated by centrifugation. The extraction was completed by adding 2 ml of chloroform-methanol (2:1, v/v) to the aqueous phase and operating as described above. The lipid phases were pooled and filtered. The washings of water-soluble and lipid extracts were performed as indicated by Sundler et al. (9). Lipid extracts were dried under nitrogen at 30°C and stored in small amounts of petroleum ether (40 to 60°C) until use. Water-soluble extracts were freed from methanol, ethanol, and chloroform and then lyophilyzed.

Determination of Phospholipids

The separation of phospholipid classes was performed according to published procedures (10). The method has been adopted for the estimation of the content of the main phospholipid classes in glial and neuronal cells and for assessing the radioactivity level of the various phospholipid spots. Since the incorporated ethanolamine and choline at all time intervals examined were located only in the EPG and CPG, respectively, the distribution of radioactivity was determined normally by treating the dried lipid extract with HCl fumes and chromatographing the diacyl-components and the monoacyl-derivatives produced from plasmalogens by one-dimensional thin-layer chromatography (TLC) using chloroform-methanol-acetone-acetic acid-water (75:15:30:15:7.5, volume ratio) as the solvent.

Pure alkenylacyl glycerophosphorylcholine (alkenylacyl-GPC) from ox heart was used as standard for the identification of choline plasmalogens in glial and neuronal extracts.

Separation of the Water-Soluble Components

The lyophilyzed aqueous extracts were dissolved in a small amount of distilled water and chromatographed on thin layers of cellulose, as indicated by Binaglia et al. (11). The separation of GPC from PC was also achieved by using the method of Dowdall et al. (12).

Radioactivity Assays

The level of [^{14}C]ethanolamine incorporated into water-soluble precursors and phospholipid was determined as previously described (7). [^{3}H]Choline incorporation was evaluated by scraping off from plates the silica gel or cellulose layer directly into the counting vials and by adding 3 ml of water and 7 ml of Instagel (Packard, Des Plaines, Ill.). Efficiency was over 25%.

Chemical Determinations

The protein content was determined on homogenate aliquots according to Lowry et al. (13). Phospholipid P was determined in lipid extracts or TLC spots according to Ernster et al. (14).

Labeled Material

[1,2-^{14}C]ethanolamine hydrochloride, [1,2-^{14}C]ethanolamine phosphate (labeled PE), [1,2-^{14}C]choline phosphate (labeled PC) and cytidine monophosphate-(CMP-)[1,2-^{14}C]choline phosphate [labeled cytidine diphosphate-choline (CDP-choline)] were obtained from New England Nuclear Corp. (Frankfurt, West Germany). CMP-[1,2-^{14}C]ethanolamine phosphate (labeled CDP-ethanolamine) was prepared as described elsewhere (15). Its purification was performed according to Chojnaki et al. (16). [Methyl-^{3}H]choline chloride was supplied by the Radiochemical Center (Amersham, England).

RESULTS

Phospholipid Composition of Neuronal and Glial Membranes

Table 1 shows that the glial fraction contains a higher amount of phospholipid on a protein basis when compared to neurons. In the rabbit brain, the glial versus

TABLE 1. *Phospholipid composition of neuronal and glial membranes*[a]

Lipid	Rabbit (present data)			Rat (17)			Chicken (18)		
	Neurons	Glia	Glial/ neuronal ratio	Neurons	Glia	Glial/ neuronal ratio	Neurons	Glia	Glial/ neuronal ratio
Phospholipid P[b]	256 ± 54 (10)	519 ± 98 (12)	2.04	387	700	1.81	346	592	1.71
Diacyl-GPC	49.5 (10)	37.8 (12)		37.0	37.0		41.6[c]	40.0[c]	
Alkenylacyl-GPC	—	1.5 (12)		1.6	1.6		—	—	
Diacyl-GPE	16.8 (10)	18.7 (12)		17.1	19.0		31.0[d]	32.0[d]	
Alkenylacyl-GPE	14.9 (10)	19.2 (12)		13.9	14.2		—	—	
Diacyl-GPS	7.2 (10)	11.1 (12)		9.8	9.9		8.1	8.8	
Phosphatidylinositol	3.9 (10)	4.8 (12)		2.9	3.0		4.9	4.8	
Sphingomyelin	4.9 (10)	6.7 (12)		5.6	5.2		8.1	9.3	

[a] Data for phospholipid classes as the percentage of phospholipid P.
[b] Mean data, as nanomoles per milligram protein ± SD values. Number of experiments in parentheses.
[c] Plus alkenylacyl-GPC.
[d] Plus alkenylacyl-GPE.

TABLE 2. *Diacyl and alkenylacyl levels of
EPG and CPG Classes in neurons and glia*

Lipid	Neurons	Glia
Diacyl-GPE	43 (10)	97 (12)
Alkenylacyl-GPE	38 (10)	99 (12)
Diacyl-GPC	127 (10)	196 (12)
Alkenylacyl-GPC	—	8 (12)

Data expressed as nanomoles per milligram
protein. Figures in parentheses represent the
number of observations.

neuronal ratio for the total phospholipid P content is approximately two when expressed on unit protein; this value is not different from those found by Freysz et al. (17) and by Freysz and Mandel (18), for rat and chicken brain, respectively. Table 1 shows also that no relevant differences exist between glia and neurons regarding the phospholipid distribution among different classes.

In Table 2 the absolute values of alkenylacyl and diacyl derivatives of CPG and EPG in glia and neurons are reported. The levels of both diacyl and alkenylacyl derivatives are higher in glial cells than in neuronal bodies. In addition the amounts of diacyl-glycerophosphorylethanolamine (diacyl-GPE) and alkenylacyl-GPE in neurons are almost identical; the same is true for the glial population. These absolute values have been adopted throughout this work for calculation of the specific radioactivity.

Incorporation of Ethanolamine into Phospholipids of Glia and Neurons

The first experiments were performed in order to show whether the preincubation period, which is necessary to dissociate the neurons and glial cells (6, 7), would modify the true levels of radioactivity at the time of sacrifice. Labeled ethanolamine was therefore injected into four rabbits, and 14 hr after administration the animals were sacrificed and the brains pooled in the same incubation medium (6, 7). Aliquots of disrupted cerebral cortex were removed at 20, 30, 45, and 60 min of preincubation and utilized separately for the preparation of neuronal and glial fractions. The analyses of lipid and of water-soluble precursors clearly demonstrated that no significant variation in the radioactivity present in lipid and aqueous extracts occurred during the intervals of preincubation examined. A preincubation time of 30 min has been adopted in preparing the cell-enriched fractions in the present experiments.

Figures 1 and 2 show the rate of ethanolamine incorporation into diacyl-GPE and alkenylacyl-GPE, respectively, as a function of time. Both phospholipids acquired their maximal specific radioactivity in neurons 7 hr after administration. On the other hand, diacyl-GPE and alkenylacyl-GPE in glia reached the highest specific radioactivity only after 20 and 36 hr, respectively. The data indicate also

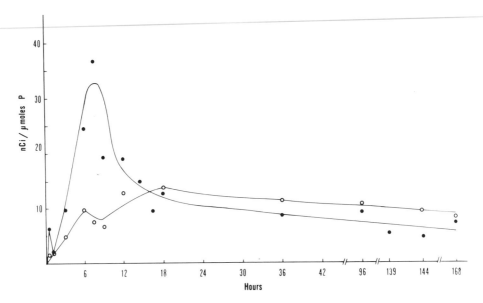

FIG. 1. Specific radioactivity of neuronal and glial diacyl-GPE at various times after the intraventricular injection of [1,2-14C]ethanolamine. Results are expressed as nanocuries incorporated per μmoles diacyl-GPE. Each point represents the mean value of three independent observations. SEM for each point was about 10 to 20%. ●—●, Neurons; ○—○, glia.

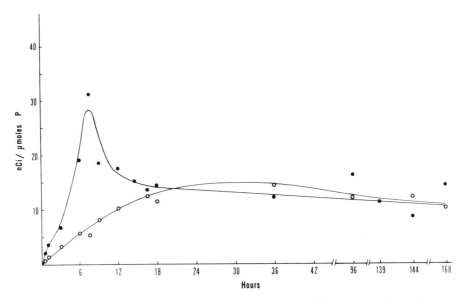

FIG. 2. Specific radioactivity of neuronal and glial alkenylacyl-GPE at various times after the intraventricular injection of [1,2-14C]ethanolamine. Results are expressed as nanocuries incorporated per μmoles alkenylacyl-GPE. See Fig. 1 for experimental details and symbols.

that in both cases the radioactivity is lost more rapidly from neuronal than from glial lipids. The glial cells possess higher activity than neurons for both EPGs 168 hr after ethanolamine administration. At that time the radioactivity of both lipids was localized only at the base level.

Incorporation of Ethanolamine into Water-Soluble Compounds

Figure 3 shows the rates of ethanolamine conversion into PE and CDP-ethanolamine of neuronal and glial cells. The highest incorporation rates for PE and CDP-ethanolamine were reached in both cells approximately 6 hr after administration. At that time, the radioactivity content of glial PE was about fourfold the value of neuronal PE on a unit protein basis, whereas that of CDP-ethanolamine was rather similar in both cell populations. The PE/CDP-ethanolamine ratio therefore was considerably lower in neurons than in glia at the maximum level of incorporating activity. At any time interval considered, no free GPE was detected in the aqueous extract; except for small amounts of free ethanolamine, no other labeled water-soluble compound was detected other than PE and CDP-ethanolamine.

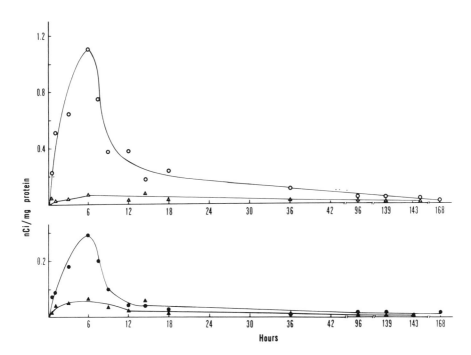

FIG. 3. Conversion of radioactive ethanolamine into PE and CDP-ethanolamine of separated neuronal and glial cells of the rabbit cerebral cortex. Results are expressed as nanocuries incorporated per mg protein. See Fig. 1 for experimental details. ●—●, neuronal PE; ▲—▲, neuronal CDP-ethanolamine; ○—○, glial PE; △—△, glial CDP-ethanolamine.

Incorporation of Choline into the Phospholipids of Glia and Neurons

The possible influence of the preincubation procedure used to prepare neuronal and glial fractions (6, 7) was first evaluated following the experimental design described above. The distribution of radioactivity of CPG acquired by both neurons and glia 4 hr after injection was similar at the various periods of preincubation adopted, suggesting no influence of the preincubation procedure, and only trace amounts of GPC were detected upon preincubation.

Figures 4 and 5 show the turnovers of diacyl-GPC and alkenylacyl-GPC, respectively. The choline incorporation was very fast in neurons for both phosphoglycerides; maximal specific radioactivity was detected within 150 min of isotope administration. In addition, a very rapid decrease of labeling occurred soon after maximal values were reached. In some experiments this decrease was even more pronounced (see Figs. 4 and 5).

At all time intervals examined, the neuronal choline plasmalogen displayed a higher specific radioactivity than the corresponding diacyl derivative. For this calculation we have assumed, according to Freysz et al. (17), that the percent of alkenylacyl-GPC content in neurons was not dissimilar from that in glia. The specific radioactivity of choline plasmalogen reached another maximal value 12 hr after administration (Fig. 5).

Both diacyl-GPC (Fig. 4) and alkenylacyl-GPC (Fig. 5) from glial cells showed a similar pattern, but the absolute levels of incorporation of choline into the two

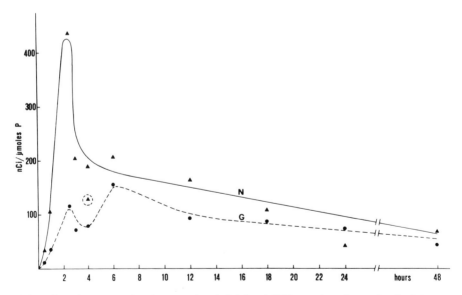

FIG. 4. Specific radioactivity of neuronal and glial diacyl-GPC at various times after the intraventricular injection of [methyl-3H]choline. Results expressed as nanocuries incorporated per μmoles diacyl-GPC. See Fig. 1 for experimental details. ▲—▲, Neurons; ●—●, glial.

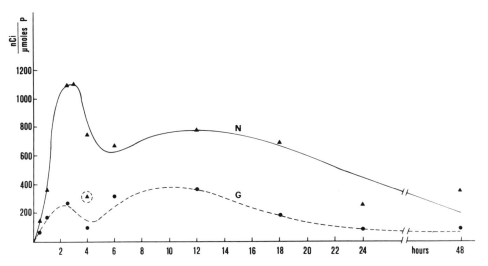

FIG. 5. Specific radioactivity of neuronal and glial alkenylacyl-GPC at various times after the intraventricular injection of [methyl-³H]choline. Results expressed as nanocuries incorporated per μmoles alkenylacyl-GPC. See Fig. 1 for experimental details. ▲—▲, Neurons; ●—●, glia.

lipids were reached at 6 hr and 10 hr, respectively. At all time intervals considered, the glial choline plasmalogen displayed a higher specific radioactivity than the corresponding diacyl derivative.

Labeled choline was also incorporated into the sphingomyelin moiety in both cell populations, although measurable levels of radioactivity could be detected within only 2 hr of administration. A small but continuous increase of the specific radioactivity of this lipid was observed up to 48 hr after choline injection.

Incorporation of Choline into Water-Soluble Compounds

Labeled choline was incorporated into PC and CDP-choline in both glial and neuronal cells. As previously reported in the case of ethanolamine incorporation (Fig. 3), higher levels of radioactivity were found in glial PC as compared to neuronal PC, and the maximal activity was detected 2 hr after injection in both cells. Similar levels of radioactivity in CDP-choline were reached in the two cell fractions, as a consequence of which the ratio PC/CDP-choline was always higher in glia than in neurons.

DISCUSSION

The results of the present work clearly show that the glial cells of the rabbit brain possess nearly a double amount of total phospholipid per milligram of protein as compared to the neuronal cells. The glial/neuronal ratio becomes even higher if the phospholipid content is referred to milligrams of DNA. Data in this connection have been reported by Freysz and Mandel (18), who found that the glial/neuronal ratio reaches values of approximately 5 when referred to DNA content of cells. Our preliminary results seem to confirm these data. By considering, in this connection, the DNA content as an approximate index of the cell number, we may indicate that glial cells contain an even higher amount of phospholipid than neurons.

The data in Table 1 show that the glial and neuronal fractions possess almost similar phospholipid composition. The fatty acid pattern of the major phospholipid classes, determined by Hamberger and Svennerholm (19), does not differ in the two cell populations. To explain the noticeable functional variations at the membrane level of glial and neuronal cells, it is probably necessary to investigate more deeply the fine arrangement of lipid and protein moieties in the respective membranes.

The pathway for ethanolamine and choline incorporation into cerebral phospholipids *in vivo* was examined by Ansell and Spanner (20, 21). Both bases were quickly phosphorylated and then transferred to lipid acceptors by means of CDP intermediates. The present work has indicated that the synthesis of CPG and EPG in glial and neuronal cells follows the same pathway described for the whole brain.

The rates of incorporation of the labeled bases into lipid is higher in the nerve cells than in glia. This result does not seem to be due to different rates of penetration of the bases into the cells, since the labeling of the water-soluble precursors, PE and PC, was faster in glia than in neurons. The accumulation of these precursors in the glial cells may be explained by the existence of a more pronounced limiting step at the level of the cytidylyltransferase enzyme. This assumption is supported by the results of Binaglia et al. (7), who found that neurons were more active than glia in synthesizing diacyl-GPC and diacyl-GPE from labeled PC and PE, respectively. In contrast, Fig. 3 shows a very rapid decrease of the labeling of glial PE at maximal activity, the same result having been obtained for glial PC. As no parallel increase of the labeling of glial EPG and CPG was observed at this stage, a possible transfer of labeled precursors from glia to neurons may also be hypothesized.

The specific radioactivity of EPG and CPG of the neuronal cells undergo a rapid decrease after maximal values were reached. At least three possible different utilizations of these newly synthesized phospholipids may occur. Recent work by Woelk et al. (8) has demonstrated that the neuronal membranes possess a higher phospholipase A activity than the glial membranes; this could be responsible for the rapid decrease of the specific radioactivity of CPG and EPG in neurons. The action of the neuronal phospholipases, however, does not seem to explain completely the rapid disappearance of lipids from the neuronal structures, since no accumulation of degradation products was observed.

Since mainly neuronal perikarya were obtained following our preparation procedure (see ref. 2), the transport of intact phospholipids by the axoplasmic flow down to the nerve endings may also be considered as one of the possible explanations for the decrease of CPG and EPG labeling. This hypothesis is partially supported by the findings of Abdel-Latif and Smith (22).

The finding that the labeling of glial phospholipids parallels the decrease of the specific radioactivity in neurons suggests a third possibility for explaining the results of the present work, namely that a transfer of intact phospholipid molecules occurs from neurons to glial cells.

SUMMARY

The phospholipid content per protein unit of glial cells from adult rabbit brain is nearly twofold that of the neuronal cells. If related to DNA content this difference is even higher. The phospholipid composition of glial and neuronal membranes is, however, not very different.

The turnover of neuronal and glial EPGs and CPGs was investigated *in vivo* following the intraventricular injection of either labeled ethanolamine or choline. Maximal values of choline and ethanolamine incorporation into the correspondent neuronal phospholipids were reached after 150 min and 7 hr, respectively, whereas highest specific activities for the correspondent glial lipids occurred much later.

ACKNOWLEDGMENTS

This investigation was aided by Research Grant 73.00730.04 from the Consiglio Nazionale delle Ricerche, Rome. Drs. G. L. Piccinin, E. Francescangeli, and R. Mozzi have collaborated at some phases of the work. Skillful technical assistance has been provided by Mr. A. Boila.

REFERENCES

1. Rose, S. P. R. (1967): *Biochem. J.*, 102:33.
2. Blomstrand, C., and Hamberger, A. (1969): *J. Neurochem.*, 16:1401.
3. Norton, W. T., and Poduslo, S. E. (1970): *Science*, 167:1144.
4. Sellinger, O. Z., and Azcurra, J. M. (1970): *Trans. Am. Neurochem.* 1:22.
5. Henn, F. A., Hansson, H. A., and Hamberger, A. (1972): *J. Cell Biol.*, 53:654.
6. Goracci, G., Blomstrand, C., Arienti, G., Hamberger, A., and Porcellati, G. (1973): *J. Neurochem.*, 20:1167.
7. Binaglia, L., Goracci, G., Porcellati, G., Roberti, R., and Woelk, H. (1973): *J. Neurochem.*, 21:1067.
8. Woelk, H., Goracci, G., Gaiti, A., and Porcellati, G. (1973): *Hoppe Seylers Z. Physiol. Chem.*, 354:729.
9. Sundler, R., Arvidson, G., and Åkesson, B. (1972): *Biochim. Biophys. Acta*, 280:559.
10. Radominska-Pirek, A., and Horrocks, L. A. (1972): *J. Lipid Res.*, 13:580.
11. Binaglia, L., Roberti, R., Michal, G., and Porcellati, G. (1973): *Int. J. Biochem.*, 4:597.
12. Dowdall, M. J., Barker, L. A., and Whittaker, V. P. (1972): *Biochem. J.*, 130:1081.
13. Lowry, O., Rosebrough, N. J., Farr, A. L., and Randall, R. J. (1951): *J. Biol. Chem.*, 193:265.
14. Ernster, L., Zetterstrom, R., and Lindberg, O. (1950): *Acta Chem. Scand.*, 4:942.
15. Porcellati, G., Biasion, M. G., and Pirotta, M. (1970): *Lipids*, 5:734.
16. Chojnaki, T., and Metcalfe, R. F. (1966): *Nature*, 210:947.

17. Freysz, L., Bieth, R., Judes, C., Sensenbrenner, M., Jacob, M., and Mandel, P. (1968): *J. Neurochem.,* 15:307.
18. Freysz, L., and Mandel, P. (1973): *FEBS Lett.,* 40:110.
19. Hamberger, A., and Svennerholm, L. (1971): *J. Neurochem.,* 18:1821.
20. Ansell, G. B., and Spanner, S. (1967): *J. Neurochem.,* 14:873.
21. Ansell, G. B., and Spanner, S. (1968): *Biochem. J.,* 110:201.
22. Abdel-Latif, A. A., and Smith, J. P. (1970): *Biochim. Biophys. Acta,* 218:134.

Lipids, Vol. 1, edited by R. Paoletti, G. Porcellati, and G. Jacini. Raven Press, New York © 1976.

Significance of Essential Fatty Acids in Human Nutrition

Ralph T. Holman

The Hormel Institute, University of Minnesota, Austin, Minnesota 55912

In 1929 Burr and Burr published the first report that fat was a necessary component of the diet of rats (1). They showed that a low-fat diet caused poor growth, scaly skin and tail, and early death, and also that linoleic acid prevented or cured these conditions. Immediately thereafter students and associates of Professor Burr attempted to study the role of polyunsaturated fatty acids (PUFA) in the diet of humans. Dr. Brown was fed a low-fat diet for several months, the iodine value of his serum lipids decreased slightly, and the only clinical effect he noticed was that his migraine headaches disappeared (2)! This study gave the impression that essential fatty acids (EFA) were not necessary for adult humans. We now know that his diet was not really free of EFA, and that a longer time would be required to induce EFA deficiency. This experiment created the impression that EFA were of minor importance in adult human nutrition, and this opinion persisted in the medical community until about a decade ago.

The most obvious symptom of EFA deficiency in animals is a scaly skin, and it was natural to seek a connection between dermatitis and EFA deficiency in humans. Dr. Hansen, a student of Burr and a pediatrician, showed in the 1930s that many cases of infant eczema that did not respond to the usual treatment for eczema did respond to a dietary supplement of lard (swine fat). He observed that those cases had a lower total unsaturation of the serum lipids than normal, and that when they were cured by dietary lard, the unsaturation of serum lipids was restored to normal (3). One such case is shown in Fig. 1 before and after a dietary supplement of lard.

The microscopic changes in the skin are shown in Fig. 2. A is the skin of a normal infant and B the skin of an infant fed a milk formula containing less than 0.1% of the calories as linoleic acid (4).

Hansen and his co-workers continued these studies and showed that in cases of infant eczema that responded to dietary EFA, the dienoic and tetraenoic acids of serum lipids were lower and trienoic acids were higher than in normal infants using alkaline isomerization as the method of analysis.

In 1959 our laboratory began a series of studies in which the effects of dose level of EFA were studied, rather than simple comparison of deficient and supplemented animals. The first of these studies involved alkaline isomerization as the analytical method; this method was later compared with gas chromatography.

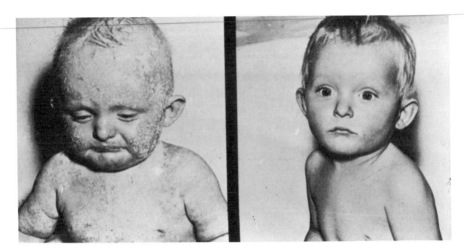

FIG. 1. A case of infant eczema refractive to usual treatments, before and after inclusion of supplement of EFA in the diet. (Courtesy of Hansen et al., ref. 3.)

The results of the two methods were found to be quite the same. In the first of these studies, the proportions of trienoic acid (5,8,11-eicosatrienoic acid) and tetraenoic acid (arachidonic acid, 5,8,11,14-eicosatetraenoic acid) in heart and liver were measured as a function of dietary intake of linoleic acid. The results showed that below 0.5% of calories of dietary linoleate, the trienoic acid was abnormally high and the tetraenoic acid was unusually low. The triene/tetraene ratio was above 1. In rats supplemented with 1% of calories of linoleic acid or more, these values were normal and the triene/tetraene ratio was below 0.4 (5). This was later found to be true for many tissues of rats, and for many species of animal.

Hansen and his group attempted to evaluate several infant formula foods containing different levels of EFA. In their study, which involved more than 400 infants, they found that skim milk formula induces EFA deficiency, judged by skin condition and by the content of trienoic and tetraenoic acids in serum lipids (4). When the data of Hansen et al. for infants were treated in the same manner as we treated the data from rats, the same relationship for rats was found (6). In Fig. 3 the triene/tetraene ratios in rat plasma and in infant serum are plotted against linoleate intake, and the two curves are nearly identical. The abrupt change in the triene/tetraene ratio occurs near 1% of calories, and this is now considered the minimum requirement for linoleic acid in both animals and man.

The same data from Hansen's infants can be used to develop an equation by which it is possible to calculate how well an individual meets the requirement for EFA (6). The relationship between log dose of linoleate and the content of dienoic, trienoic, and tetraenoic acids in serum lipids is shown in Fig. 4. These data and this equation permit estimate of linoleate intake to within about 25% of the true value. This is precise enough for the practical purpose of deciding

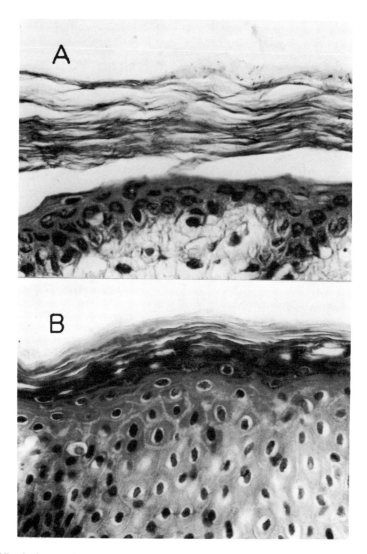

FIG. 2. Histologic examination of the skin of **(A)** a normal infant and **(B)** the skin of an infant exhibiting dermatitis from EFA deficiency. (Courtesy of Hansen et al., ref. 4.)

whether an infant is deficient (less than 1% of calories) or is in the normal range ($>$ 5% of calories).

A similar study was made of 12 young adult males (7), and the estimated equation for them is shown in Fig. 5. In this study it was not possible to feed really deficient diets, so we were unable to set the EFA requirement for adult males. We believe it is probably very close to the requirement for children.

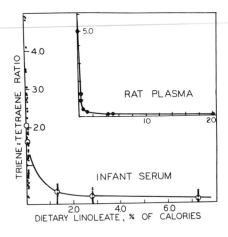

FIG. 3. Triene:tetraene ratios for serum lipids of infants and plasma lipids of rats receiving different levels of dietary linoleic acid. (Data from Holman, ref. 5, and Holman et al., ref. 6.)

Judging from our studies on rats, the EFA requirement for females may be less than the requirement for males.

If the apparent dietary linoleate, calculated as [diene − triene + tetraene] of serum, is plotted against the age of breast-fed infants, newborn infants appear to be EFA deficient, and their serum approaches the normal values for PUFA within a few days. However, if young infants are fed only cow's milk formula, the triene/tetraene ratio increases, whereas in breast-fed infants the ratio remains low, as is shown in Fig. 6 (8). This suggests that cow's milk may induce low-level EFA deficiency in infants.

There is some evidence that faulty metabolism of EFA may be a factor in some rare diseases. *Achrodermatitis enteropathica* is an infant disease involving severe

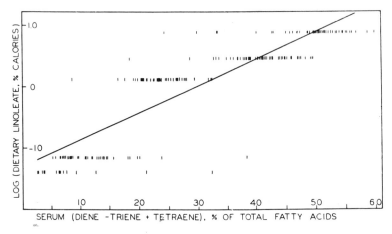

FIG. 4. Estimation equation describing the relationship between dietary intake of linoleate and content of PUFA in serum lipids of infants 2 to 4 months old. Log$_{10}$ (dietary linoleate)= −1.087 + 0.0432(diene-triene + tetraene). (Data from Holman et al., ref. 6.)

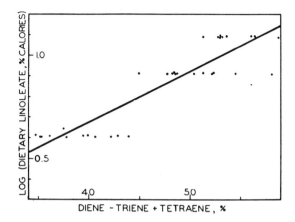

FIG. 5. Estimation equation describing the relationship between dietary intake of linoleate and content of PUFA in serum lipids of young men. Log lin. $= -0.296 + 0.024$(diene-triene + tetraene). (Data from Holman et al., ref. 7.)

dermatitis and malabsorption. In his lengthy study of this disease, Dr. Ralph Cash of Detroit has concluded that the cause is an error in absorption or metabolism of linoleic acid. He uses intravenous infusion of fat emulsion as a diagnostic tool. In Fig. 7 one of his cases is shown at the time of admission to the hospital. After a few days of intravenous fat emulsion, the dermatitis had improved significantly. Withdrawal of the fat emulsion for a few days caused the return of the dermatitis. The diagnosis was thus made that the case was *A. enteropathica,* which usually responds to a diet containing no cow's milk and daily administration of Diodoquin. After several days on this therapeutic regime, the dermatitis disappeared. The fatty acid composition of the serum lipids was analyzed by gas-liquid chromatography (GLC) at each stage of the diagnosis and treatment, and the analyses were confirmed in our laboratory. In conditions A and C the lipids had a low content of arachidonic acid and a high content of unusual fatty acids, perhaps odd-chain and branched-chain fatty acids. When the condition was

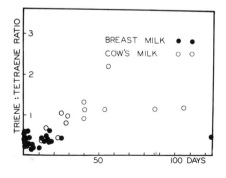

FIG. 6. Triene:tetraene ratio of serum PUFA of infants fed breast milk (●) or cow's milk (○) formula. (Data from Holman et al., ref. 8.)

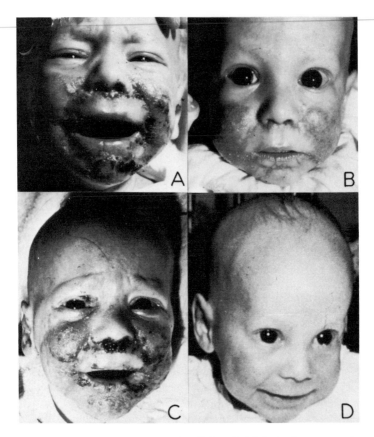

FIG. 7. A case of *A. enteropathica* **(A)** as received in hospital, **(B)** after intravenous fat emulsion, **(C)** after withdrawal of fat emulsion and restoration to usual diet, and **(D)** after Diodoquin and a diet excluding cow's milk (9).

improved by intravenous fat emulsion or by Diodoquin and no cow's milk, the arachidonic acid increased toward normal level and the unusual fatty acids decreased. The case was clearly one of abnormal lipid absorption or metabolism (9).

Dr. Whitten, of Detroit, was caring for a case of radical bowel resection and suspected the child was developing EFA deficiency. The child had experienced extreme intestinal upset shortly after birth, and exploratory surgery discovered the mesentery had rotated at birth, cutting off the blood supply and killing the intestine. Joining of the duodenum to the colon left the child with very little absorptive intestine and constant intravenous feeding was required to support life. The usual intravenous infusion consists of glucose, amino acids, minerals, and vitamins, but no fat. It is therefore an extremely EFA-deficient diet. The extensive dermatitis on the child at 3 months of age is shown in Fig. 8. Samples of serum taken serially were analyzed for fatty acid composition of the triglycerides, cho-

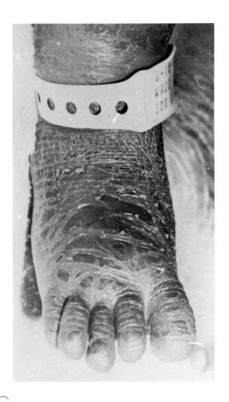

FIG. 8. Dermatitis of EFA deficiency induced in a child by prolonged intravenous feeding with a fat-free preparation. (From Paulsrud et al., ref. 10.)

lesteryl esters, and phospholipids, the results of which indicated very abnormal PUFA patterns and severe EFA deficiency (10). The triene/tetraene ratio at about 100 days of age was 16, to be compared with a ratio of 6 for the most severely deficient rat analyzed.

This observation led to a study of seven infants who for one reason or another required long-term intravenous feeding (10). They uniformly underwent the same changes in PUFA pattern. The changes in PUFA content of serum phospholipids are shown for the seven infants in Fig. 9. The upward trend for $20:3\omega9$ (eicosatrienoic acid) and the downward trend for $20:4\omega6$ is the same for each individual, although the magnitude of the change is different for each. For those fortunate children who could later be given normal food, the abnormal patterns changed to normal, indicating that these biochemical changes are reversible.

The severe case mentioned above is shown in Fig. 9 with double circles. This child lived for 4.5 months, and was fed continuously by intravenous infusion. At autopsy samples of tissue were taken, the fatty acid compositions of tissue lipids were analyzed, and the results are shown in Table 1. These results show that in all tissues the fatty acid characteristic of EFA deficiency, 5,8,11-eicosatrienoic acid, is present in very high proportions, and arachidonic acid is present in abnormally low amounts. This case is the most completely described and the most severe case of EFA deficiency in a human thus far reported. The similarity in

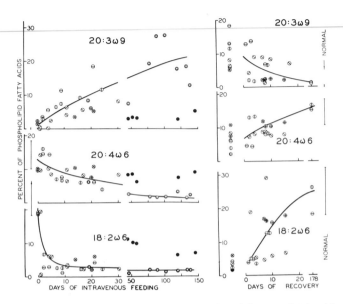

FIG. 9. Fatty acid composition of serum phospholipids from infants maintained by prolonged intravenous alimentation with a fat-free preparation **(left)** followed by a recover period **(right)** during which normal food was given. (From Paulsrud et al., ref. 10.)

fatty acid composition of tissue lipids to those from EFA-deficient animals is striking, and is a proof that EFA deficiency can be induced in man.

A report of a similar case in a 44-year-old man has been made from Australia by Collins et al. (11). They followed the content of 20:3 and 20:4 in serum lipids during two cycles of fat-free intravenous feeding followed by administration of a fat emulsion. Their results are shown in Fig. 10. When fat-free intravenous fluid was given, 20:3 was high and 20:4 was low. When fat emulsion containing

TABLE 1. *Content of polyunsaturated fatty acids in phosphatidylcholine fatty acids from tissues of an EFA-deficient infant*

Tissue	18:2	20:3ω9	20:4ω6	Triene/ tetraene
Liver	1.1	9.9	2.0	5.0
Kidney	0.7	7.0	2.6	2.7
Pancreas	0.6	4.8	0.8	6.0
Adrenal	0.7	11.0	2.8	3.9
Colon	0.9	4.8	1.9	2.5
Ventricle	0.3	7.8	3.1	2.5
Spinal cord	0.6	1.1	1.0	1.1
Serum	1.8	13.1	2.2	6.0

Expressed as percentage of total fatty acids.
(Data from Paulsrud et al., ref. 10.)

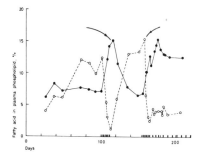

FIG. 10. Response of 20:3ω9 (○) and 20:4ω 6 (●) to intravenous feeding with fat-free preparation and with fat emulsion in a 44-year-old male. Administration of fat emulsion is indicated by bars along the abscissa. (From Collins et al., ref. 11.)

linoleate was given, 20:3 dropped to normal, and 20:4 increased to normal. It is clear that intravenous fat emulsion corrects the biochemical deficiency of EFA induced by fat-free intravenous alimentation.

EFA has value in human physiology beyond prevention of dermatitis. EFA is required in large amounts during periods of growth of tissue. Hence EFA deficiency can be induced in infants and young animals much more easily than in adults. Other conditions involving proliferation of tissue might be thus expected to require EFA in an increased amount. Indeed Caldwell et al. (12) found wound healing to be defective in infants who have been maintained for long periods of time on fat-free intravenous nutrient. Figure 11A shows the surgical wounds of a child who underwent repeated surgery for bowel resection and whose wounds did not heal during the period it was fed intravenously a fat-free preparation. Figure 11B shows the same child after it had been fed for some weeks with fat emulsion containing EFA. Thus, EFA is required for proper wound healing. There is also reason to suspect that EFA deficiency makes bacterial invasion

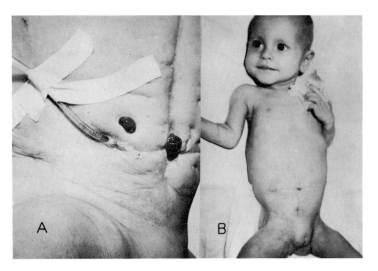

FIG. 11. Lack of wound healing in condition of EFA deficiency **(A)** and healed wounds after administering intravenous fat emulsion **(B).** (From Caldwell et al., ref. 12.)

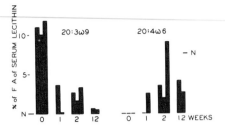

FIG. 12. Effect of cutaneous application of sunflower seed oil upon fatty acids of serum lipids. **(Left)** 20:3ω9, **(right)** 20:4ω6. From Press et al., ref. 13.

easier, and that this tendency toward infection is increased in long-term intravenous feeding with fat-free preparations.

Recently a group of British workers have demonstrated that EFA deficiency that had been induced by long-term intravenous feeding with a fat-free preparation, could be relieved by absorption of unsaturated oils through the skin (13). Press et al. (13) showed, in fact, that rubbing sunflower seed oil into the skin caused sufficient absorption to not only relieve the dermatitis of deficiency, but to reverse the abnormal pattern of PUFA in serum lipids. Their results (Fig. 12) show that within a few weeks, the daily application of oil to the skin caused a significant reduction of 20:3ω9 and an increase in 20:4ω6 in serum lecithin. Thus the oil had been absorbed in sufficient amounts to cure the biochemical deficiency.

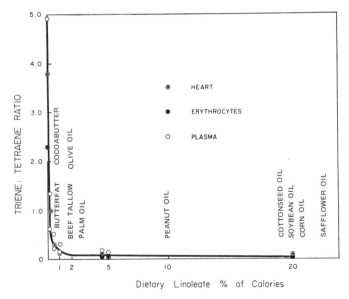

FIG. 13. Dietary linoleate provided by 40% of calories of some food fats superimposed over the curve showing EFA requirement. (From Holman, ref. 5.)

It may not then be necessary to administer unsaturated fat by the hazardous intravenous route.

All of this evidence shows that EFA are necessary for human nutrition and metabolism, and it is then evident that adequate amounts of EFA must be provided by the diet to insure health. The minimum requirement is in the order of 1% of calories of linoleic acid, but some individuals may require considerably more. The requirement for optimum nutrition has not been defined (14), but 5 to 10% of calories of linoleate appears to be reasonable. High levels of polyunsaturated acids have some adverse effects, probably related to the exhaustion of limited tocopherol. Very high levels of PUFA in rats cause dermal symptoms to appear and growth to diminish. Hence care should be exercised not to consume too high proportions of PUFA. If a little is good, it does not necessarily follow that a lot is better.

How much and what kind of fat should be included in the diet to assure an adequate intake of EFA is often questioned. The American population consumes approximately 40% of its calories as fat. If all of this fat were consumed as butter fat, the intake of linoleic acid would be 0.5 to 1% of calories (Fig. 13). If, on the other hand, all the dietary fat were sunflower seed oil, corn oil, or cottonseed oil, about 20% of calories of linoleic acid would be provided. Other sources of dietary fat range in between these limits. No population, no culture, and even no individual consumes its fat all from one source, so it is unlikely that a mixed diet could result in an EFA deficiency in humans. Only in those cultures that depend heavily on cow's milk fat and body fats of ruminant animals as a major source of calories is it possible that even marginal EFA deficiency could arise in normal persons. Some individuals may have greater requirements than others because of inherited errors in metabolism, and such cases should be investigated to learn whether high intakes of EFA would help correct their deficiencies. For the great majority of the rest of us, the danger of EFA deficiency is minimal. We need only extend the principles of balanced nutrition to include a balanced variety of dietary fats in the human diet to assure that the requirement for essential fatty acids will be met.

REFERENCES

1. Burr, G. O., and Burr, M. M. (1929): *J. Biol. Chem.*, 82:345.
2. Brown, W. R., Hansen, A. E., Burr, G. O., and McQuarrie, I. (1938): *J. Nutr.*, 16:511.
3. Hansen, A. E., Knott, E. M., Wiese, H. F., Shaperman, E., and McQuarrie, I. (1947): *Am. J. Dis. Child.*, 73:1.
4. Hansen, A. E., Wiese, H. F., Boelsche, A. N., Haggard, M. E., Adam, D. J. D., and Davis, H. (1963): *Pediatrics*, 31:Suppl. 1, pt 2, 171.
5. Holman, R. T. (1960): *J. Nutr.*, 70:405.
6. Holman, R. T., Caster, W. O., and Wiese, H. F. (1964): *Am. J. Clin. Nutr.*, 14:70.
7. Holman, R. T., Caster, W. O., and Wiese, H. F. (1964): *Am. J. Clin. Nutr.*, 14:193.
8. Holman, R. T., Hayes, H. W., Rinne, A., and Söderhjelm, L. (1965): *Acta Paediat. Scand.*, 54:573.
9. Cash, R. and Berger, C. K. (1969): *J. Pediat.*, 74:717.

10. Paulsrud, J. R., Whitten, C. F., Stewart, S. E., and Holman, R. T. (1972): *Am. J. Clin. Nutr.,* 25:897.
11. Collins, F. D., Sinclair, A. J., Royle, J. P., Coats, D. A., Maynard, A. T., and Leonard, R. F. (1971): *Nutr. Metab.,* 13:150.
12. Caldwell, M. D., Jonsson, H. T., and Othersen, H. B. (1972): *Pediatrics,* 81:894.
13. Press, M., Hartop, P. J., and Prottey, C. (1974): *Lancet,* 1:594.
14. Holman, R. T. (1971): *Progress in the Chemistry of Fats and Other Lipids, Vol. IX,* p. 672. Pergamon Press, Oxford.

ACKNOWLEDGMENTS

Experimental work included in this chapter was supported in part by the National Institutes of Health (AM 04524) and by The Hormel Foundation.

Lipids, Vol. 1, edited by R. Paoletti, G. Porcellati, and G. Jacini. Raven Press, New York © 1976.

Gangliosides and Glycoproteins in Brain of Rats Fed on Different Fats

B. Berra, M. Ciampa, G. Debernardi, M. Manto, and V. Zambotti

Department of Biological Chemistry, University of Milan, School of Medicine, 20133 Milan, Italy

It is well established that nutritional factors can influence some biochemical parameters in the central nervous system. In undernourished animals the amount of cholesterol (1), phospholipids (1), and cerebrosides (2) is lower than in normals, the synthesis of sulfatides is diminished (3), and the process of myelination is delayed (4). The absence of a low content of essential fatty acids in the diet may influence the normal brain development (5). On the contrary, very little is known of the influence of nutritional factors on the amount and metabolism of other important brain compounds, such as gangliosides and sialoglycoproteins.

Gangliosides are mainly located in neuronal membranes, for which they are considered markers; their amount and the relative ratio of the different species of gangliosides are correlated to some important morphologic changes that take place in the central nervous system during development (6, 7). Sialoglycoproteins are essential components of all brain membranes and are thought to play a role in the differentiation, migration, and adhesion mechanisms of the nerve cells.

Both of these groups of substances contain sialic acid, which seems to be the active component of the molecule. Both the number and the position of the sialic residues can modify the chemicophysical properties of the molecules, with functional implications. In this sense, a regulation mechanism should be a repetition of splitting off and reincorporation of sialic acid, the first reaction being catalyzed by the enzyme neuraminidase (NANA, *N*-acetylneuraminylhydrolase, EC 3.2.1.18).

This work investigates the influence of olive oil on the amount and the patterns of different brain gangliosides, their fatty acid composition, the amount of sialoglycoproteins, and the activity of neuraminidase. These parameters were chosen because of their functional importance, as above stated. The same parameters were determined in animals fed on diets containing an equal amount of other fats (sunflower oil, tallow, hydrogenated fat) and in animals fed on a standard diet, considered as controls.

MATERIALS AND METHODS

Animals and Diets

Sprague-Dawley rats were fed *ad libitum* 10 days before mating, with four different diets, of which the basic composition was fat-free casein, 18%; sucrose,

TABLE 1. *Fatty acid distribution of different diets*

Fatty acid	Olive oil (%)	Tallow (%)	Sunflower oil (%)	Saturated fat (%)	Normal rat diet (%)
10:0	—	1.1	—	—	—
12:0	tr.	0.6	—	—	0.1
14:0	0.2	5.7	0.3	3.3	1.5
15:0	—	0.7	0.1	0.9	0.2
15:1	—	0.3	—	0.2	—
16:0	12.2	32.4	7.9	31.6	17.9
16:1	1.2	5.0	0.3	0.8	2.5
17:0	—	1.4	1.0	2.5	0.4
17:1	0.2	0.9	—	0.2	0.4
18:0	2.9	2.2	5.3	56.5	5.6
18:1	74.6	44.0	22.2	1.1	28.0
18:2	6.3	2.5	59.7	3.0	32.6
20:0	0.4	0.2	0.4	0.6	0.3
18:3	1.1	0.8	0.6	—	4.1
20:2	0.3	0.5	0.3	0.2	0.5
20:3	—	—	0.9	0.2	0.2
20:3	—	—	—	—	—
20:4	—	—	—	—	0.4
20:5 (*n*-3)	—	—	—	—	1.8
22:3	—	0.3	—	0.3	0.3
22:4	—	0.3	—	—	0.3
22:5 (*n*-3)	—	0.7	—	—	0.2
22:6	—	—	—	—	2.2

Data are shown in percentage values.

50%; starch, 12%; α-cellulose, 4%, salt mixture, 4%, choline chloride, 1%, vitamin mixture, 1%. To each diet 10% of olive oil, sunflower oil, tallow, and saturated fat (obtained by hydrogenation of tallow) was added, respectively.

A control group of animals was fed on a standard diet with the following basic composition: protein, 26%, fat, 5%, α-cellulose, 3%, sucrose, 42%, mineral, 4%, vitamin mixture, 1%. The fatty acid composition of the diets is reported in Table 1.

The animals were fed with the different diets during pregnancy and lactation: the same diets were used for the weaned animals from the 21st day after birth.

The number of rats in all litters was reduced, at birth, to six. The animals were sacrificed at 1, 20, 40, and 100 days of age; for the analysis three litters for each treatment and for each time interval were used. Brains from a single litter were pooled and analyzed in duplicate.

Total brain of animals sacrificed on the 1st day was analyzed; for the animals sacrificed on the 20th, 40th, and 100th day, investigations were performed only on the encephalon.

Reagents and Methods

All reagents of analytical grade were used; when used in gas-liquid chromatographic (GLC) analysis they were redistilled before use. Gangliosides were extracted, fractionated, and determined after extraction with tetrahydrofuran (8). The single gangliosides were methanolyzed with boron trifluoride, and fatty acid methyl esters purified by thin-layer chromatography (TLC) (9). The following conditions for the GLC analysis were used: Gas-chromatograph C. Erba mod. Fractovap 2400 V; 2-m glass column packed with 10% Lac 728 coated on CWLA 60 to 80 mesh; temperature 186°C; carrier gas: N_2.

The defatted residue obtained after the extraction of gangliosides was homogenized and hydrolized with 0.1-N H_2SO_4 at 80°C for 30 min. The homogenate was centrifuged; the supernatant was carefully collected, and the pellet was hydrolyzed again in the same conditions and centrifuged. This operation was repeated twice. All the supernatants were pooled and the final volume adjusted to 50 ml. An aliquot of 20 ml was chromatographed on a Dowex 2 × 8 (acetate form) column (0.8 × 6 cm) and the NANA was eluted with 6 ml of 2-M acetate buffer, pH 4.6. On the eluted solution the amount of NANA was determined with thiobarbituric acid (10). The total and specific activity of particulate neuraminidase was determined as previously reported (11).

RESULTS AND DISCUSSION

The total amount of ganglioside and glycoprotein NANA, expressed as micromoles per gram dry brain weight in animals fed on the normal diet are reported in Fig. 1. For the ganglioside-bound NANA, there is a sharp increase from the 1st to the 20th day after birth, followed by a smaller increase until the 40th day, when the maximum amount was found; at 100 days of age the values are slightly

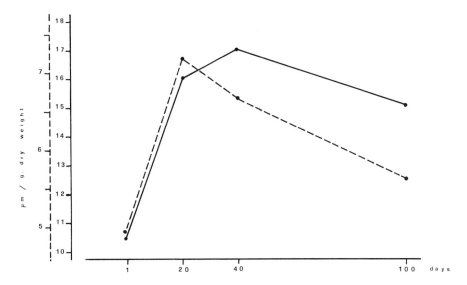

FIG. 1. Content of ganglioside and glycoprotein NANA (expressed as micromoles per gram dry weight) in rats fed a normal diet at various ages (**dotted line,** glycoprotein NANA; **black line,** ganglioside NANA).

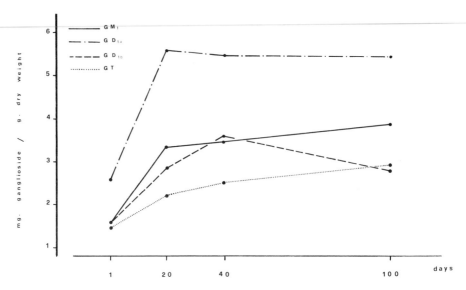

FIG. 2. Patterns of the four major gangliosides (expressed as milligrams per gram dry weight) in rats fed on normal diet at various ages. — GM_1, —.— GD_{1a}, — — — GD_{1b}, . . . GT.

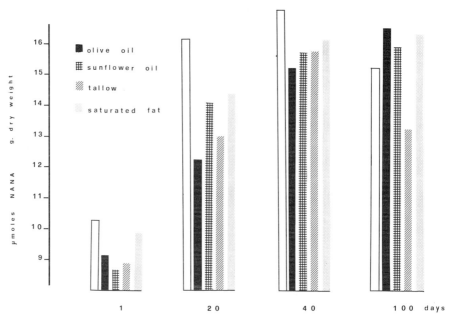

FIG. 3. Content of total gangliosides (expressed as micromoles of lipid-bound NANA per gram of dry weight) in rats fed on diets with equal amounts of different fats (void bars indicate the values of normally fed rats as reported in Fig. 1).

lower than the values found at 20 days. A parallel increase between birth and 20 days was found for glycoprotein NANA, with the difference that the maximum value was reached at 20 days; then the amount steadily decreases.

The patterns of the four major gangliosides, GM_1, GD_{1a}, GD_{1b}, GT in control animals is reported in Fig. 2. Ganglioside GD_{1a} increases from birth until 20 days of age and then remains practically constant. Gangliosides GM_1 and GT increase from birth to 100 days (with a maximum increasing rate between birth and the 20th day), whereas ganglioside GD_{1b} increases from birth to 40 days, when there is a maximum value; then there is a decrease; at 100 days we found quite the same values as at 20 days.

The content of ganglioside NANA in rats fed on four semisynthetic diets is reported in Fig. 3. There are some differences among the four groups of animals, particularly the high amount of NANA at birth in the animals fed on saturated fat and the low amount at 100 days in the animals fed on tallow. In comparison with control animals, the total amount of lipid-bound NANA for treated rats is lower at birth and remains lower at 20 and 40 days. At 100 days the value for animals fed on olive oil, sunflower oil, and saturated fat is slightly higher than normal. Only tallow-fed animals show a decrease, like normals, even with a more rapid rate.

The pattern of glycoprotein NANA content, as reported in Fig. 4, shows more evident differences among the four groups of treated animals; for this parameter too, the highest amount was found at birth for the animals fed on saturated fat,

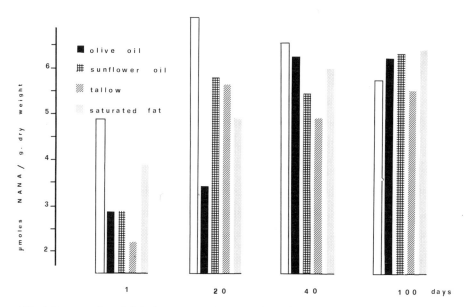

FIG. 4. Content of total sialoglycoproteins (expressed as micromoles of NANA per gram of dry weight) in rats fed on diets with equal amounts of different fats and control diets (void bars indicate the value of normally fed rats as reported in Fig. 1).

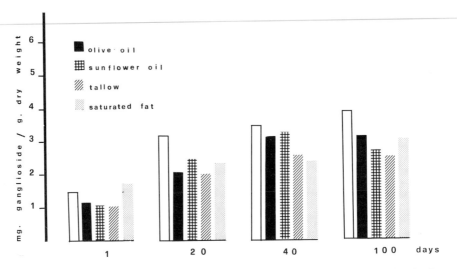

FIG. 5. Content of ganglioside GM$_1$ (expressed as milligrams per gram dry weight) in the four groups of animals fed on semisynthetic and control diets (void bars indicate the value of normally fed rats as reported in Fig. 2).

whereas a significantly lower value was observed for animals fed on tallow. Olive oil-fed animals show a very slow increase from birth to 20 days. In comparison with the data obtained with the control animals, there are again lower values for the treated animals at birth, 20, and 40 days. At 100 days after birth, animals

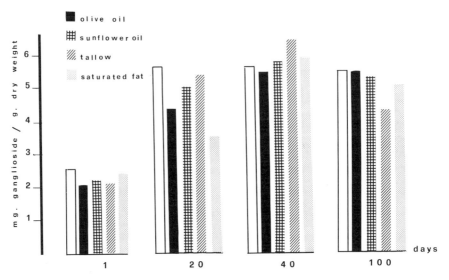

FIG. 6. Content of ganglioside GD$_{1a}$ (expressed as milligrams per gram dry weight) in the four groups of animals fed on semisynthetic and control diets (void bars indicate the value of normally fed rats as reported in Fig. 2).

fed on olive oil, sunflower oil, and saturated fat show, as observed for lipid-bound NANA, slightly higher value than normal.

The distribution of ganglioside GM_1 in the four groups of treated rats is reported in Fig. 5; no significant difference was noted between the four groups, except for the higher amount at birth in the animals fed on saturated fat. In contrast with controls, particularly evident are the lower values at 20 days and a decrease at 100 days in the animals fed on tallow.

The distribution of ganglioside GD_{1a} is reported in Fig. 6. There is some difference between the animals fed on saturated fat and the other three groups at 20 days of age; in comparison with control rats, it can be observed that in all four groups the maximal values attained shift from the 20th to the 40th day; this is particularly evident for animals fed on saturated fat. No significant differences were noted between the four groups for gangliosides GD_{1b} (Fig. 7) and GT (Fig. 8). For ganglioside GT the four groups of treated animals showed higher values at birth than normally fed rats.

Brain gangliosides contain mainly stearic acid in their molecule (85 to 90% of the total fatty acids). As the different composition of the diet modifies the fatty acid pattern in other structural lipids as phospholipids (12), we have also investigated this parameter under our experimental conditions. The pattern of fatty acid in the single gangliosides is unaffected by the different ratio of dietary fatty acids.

Practically no significant difference was found in the total and specific activity of the neuraminidase, at least at the ages examined (20, 40, and 100 days after birth); unfortunately, we were unable to obtain sufficient material to investigate

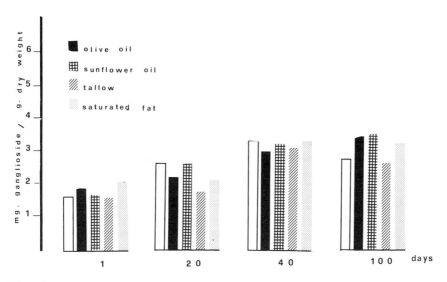

FIG. 7. Content of ganglioside GD_{1b} (expressed as milligrams per gram dry weight) in the four groups of animals fed on semisynthetic and control diets (void bars indicate the value of normally fed rats as reported in Fig. 2).

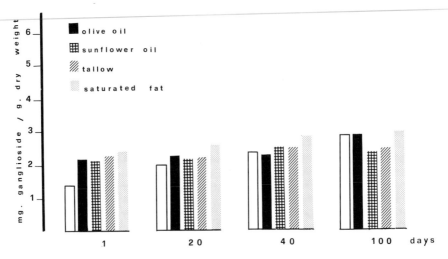

FIG. 8. Content of ganglioside GT (expressed as milligrams per gram dry weight) in the four groups of animals fed on semisynthetic and control diets (void bars indicate the value of normally fed rats as reported in Fig. 2).

this activity at the 1st day after birth, in which we found significant differences of the lipid-bound NANA and of the pattern of the single gangliosides.

In conclusion our data on ganglioside and sialoglycoprotein content suggest that all the investigated diets induce a delay in reaching the value obtained with the normal diet. This is more evident for sialoglycoproteins than for gangliosides. We still do not know if this is because of the different composition of the control diet, which contains a lower amount of fat with a different percentage of fatty acids.

The main differences observed on studying the patterns of the four major gangliosides and the total amount of glycoprotein and lipid-bound NANA were found in the period between birth and the 20th day. It must be pointed out that this period corresponds, from a morphologic point of view, to the multiplication of glial cells and microneurons, the outgrowth of dendrites and axons, and the establishment of neuronal connections with a corresponding first phase of rapid increase of ganglioside and glycoprotein concentration. The physiologic implications of these differences are still obscure.

SUMMARY

The total amount of gangliosides, their pattern and their fatty acid composition, the activity of neuraminidase, and the total sialoglycoprotein content were investigated in brains of rats of different ages fed equal amount of different fats, in comparison with animals fed on a normal standard diet. The results indicate that all diets influence the amount and the ratio of investigated sialocompounds, mainly during the period between birth and the 20th day.

The amount of these compounds is not influenced by the content of essential fatty acids in the diets; the fatty acid distribution of brain gangliosides is uneffected by the different fat composition.

REFERENCES

1. Dobbing, J., and Widdwson, E. M. (1965): *Brain,* 88:357.
2. Culley, W. J., and Mertz, E. T. (1965): *Proc. Soc. Exp. Biol. Med.,* 118:233.
3. Chase, H. P., Dorsey, and McKhann, G. M. (1967): *Pediatrics,* 4:551.
4. Benton, J. U., Moser, H. W., Dodge, P. R., and Carr, S. (1966): *Pediatrics,* 38:801.
5. Paoletti, R., and Galli, C. (1972): *Lipids, Malnutrition and Developing Brain,* p. 121. Elsevier, Amsterdam.
6. Vanier, M. T., Holm, M., Ohman, R., and Svennerholm, L. (1971): *J. Neurochem.,* 18:581.
7. Herschkowitz, N. (1973): *Proc. 2nd Giornate di Studio Plasmon,* Milano, p. 217.
8. Tettamanti, G., Bonali, F., Marchesini, S., and Zambotti, V. (1973): *Biochim. Biophys. Acta,* 296:160.
9. Rouser, G., and Yamamoto, A. (1972): *J. Neurochem,* 19:2697.
10. Warren, L. (1959): *J. Biol. Chem.,* 234:1971.
11. Tettamanti, G., Venerando, B., Preti, A., Lombardo, A., and Zambotti, V. (1972): *Adv. Exp. Med. Biol.,* 25:161.
12. Galli, C., Trzeciak, H. O., and Paoletti, R. (1971): *Biochim. Biophys. Acta,* 248:449.

Lipids, Vol. 1: Biochemistry, edited by R. Paoletti, G. Porcellati, and G. Jacini. Raven Press,
New York © 1976.

Comparative Effects of Olive Oil and Other Edible Fats on Brain Structural Lipids During Development

C. Galli, C. Spagnuolo, E. Agradi, and R. Paoletti

Institute of Pharmacology and Pharmacognosy, University of Milan, 20129 Milan, Italy

The influence of the composition of dietary fat on the fatty acids of structural lipids in the developing brain has been shown in several laboratories (1–4) including our own (5–8). Attention has been paid especially to the role of essential fatty acids (EFA) and to the changes of the fatty acid composition of brain structural lipids induced by the deficiency of these essential compounds.

From a biochemical point of view, a reduction of EFA intake results mainly in accumulation in tissue lipids of fatty acids with three double bonds deriving from oleic acid (trienes) and reduction of polyunsaturated fatty acids with four double bonds (tetraenes) derived from linoleic acid. The consequent rise of the triene/tetraene ratio in tissue lipids is considered a biochemical index of EFA deficiency at the tissue level (9). Brain has been shown to be sensitive to dietary EFA deficiency, although the biochemical effects are less pronounced and more slowly induced than those observed in other tissues.

Since brain phospholipids contain high levels of polyunsaturated fatty acids of both the linoleic (18 : 2 n-6) and linolenic (18 : 3 n-3) acid series, consideration should be given also to the levels of both these fatty acids in the diet. It has been shown, in fact, that changes in the linoleic/linolenic acid ratio in the lipid fraction of the diet modify the relative proportion of polyunsaturated fatty acids derived from these essential precursors in tissues, including brain (2, 8) and brain subcellular structures (10).

However, in spite of considerable variations in the fatty acid composition of cerebral phospholipids induced by changes in the levels and relative proportions of dietary EFA, the degree of unsaturation of these structural lipids is retained (5, 8). The homeostatic control observed in brain, in retaining the level of unsaturation of structural phospholipids, irrespective of the fatty acid composition of dietary lipids, is not observed, however, under the same dietary conditions in the liver (11).

The competitive reciprocal inhibition in the conversion of linoleic (n-6 series) and linolenic (n-3 series) acids to their longer chain more unsaturated derivatives is responsible for the marked modifications in the relative levels of n-6 and n-3 fatty acids detected in tissue lipids of animals fed on diets containing different relative proportions of EFA. More specifically variations in the relative ratio of linoleic to linolenic acid results in changes of the relative levels of the last members of the *n*-6 and *n*-3 fatty acid series (22 : 5 *n*-6 and 22 : 6 *n*-3, respectively) (2, 3, 8).

We proposed to consider the 22 : 5 *n*-6/22 : 6 *n*-3 ratio in tissue lipids as an index of relative linolenic acid deficiency (12) in analogy with the significance of the 20 : 3 *n*-9/20 : 4 n-6 ratio as an index of total EFA deficiency. It appears thus that both the total levels and relative proportions of linoleic to linolenic acid are to be considered in order to evaluate dietary supplies of EFA.

PRESENT STUDY

The present study is concerned with the effects of the administration of semi-synthetic diets, containing constant amounts of fat with varying levels and proportions of EFA, on the fatty acid composition of ethanolamine phosphoglyceride (EPG) in the developing brain, in the rat. EPG is the major, highly unsaturated phospholipid in the central nervous system. More specifically the effects of the administration of the standard diet or of semisynthetic diets containing either olive oil or other edible fats on the triene/tetraene and on the pentaene/hexaene ratios in EPG were studied in the developing brain of the rat.

EXPERIMENTAL CONDITIONS

The following experimental diets, (a) standard diet, or semisynthetic diets containing 10% of either (b) olive oil, (c) sunflower seed oil, (d) tallow, and (e) saturated fat were fed to corresponding groups of female rats of the Sprague-Dawley strain, starting 10 days before mating. Treatments were continued during pregnancy and lactation, and, after weaning, the same diets were fed to the young rats. At birth the litter size was reduced to six animals in each group. Three litters were sacrificed in each group at the ages of 1, 20, 40, and 100 days. At 1, 20, and 40 days of age, animals of both sexes were used, whereas at 100 days only male rats were analyzed. The animals were sacrificed by decapitation, brains were immediately removed and pooled in a single pool for each dietary treatment at the age of 1 day, whereas, at subsequent age intervals, only brains obtained from animals of the same litter were pooled. Thus three different pools of brains for each type of dietary treatment were used at the ages of 20, 40, and 100 days.

Lipids were extracted from the pooled samples (13). EPG was isolated and the fatty acid methyl esters of both the mono- and diacyl moieties of this compound were prepared, purified from aldehydes, and analyzed as previously described (5).

RESULTS AND DISCUSSION

The fatty acid composition of the lipid fraction in the various diets is shown in Table 1. Lipid extracted from the normal diet and the diet containing sunflower seed oil (SO) have similar polyenoic acid contents and unsaturation indexes, but SO has a much higher (*n*-6)/(*n*-3) fatty acid ratio. In contrast, olive oil (OO) and

TABLE 1. *Unsaturation levels of fatty acid methyl esters of dietary lipids*

	N	OO	SO	T	SF
Saturated	26.0	15.8	14.1	44.4	96.1
Monoenes	30.9	76.0	22.5	50.3	2.6
dienes	33.1	6.7	60.0	3.1	0.5
trienes	4.6	1.1	1.4	1.4	0.4 (*n*-9)
tetraenes	0.7	—	—	—	—
pentaenes	2.1	—	—	0.7	—
hexaenes	2.2	—	—	—	—
Polyunsaturates	42.7	7.8	61.4	5.2	0.9
Unsaturation index[a]	137	93	147	61	5
(*n*-6)/(*n*-3)	4.0	5.6	105.0	3.0	—

N, normal diet; OO, diet containing 10% olive oil; SO, diet containing 10% sunflower seed oil; T, diet containing 10% tallow; SF, diet containing 10% saturated fat.

[a] Sum of percentage of individual unsaturated fatty acid X number of double bonds.

tallow (T) have low polyenoic acid contents and lower levels of unsaturation, but the (*n*-6)/(*n*-3) fatty acid ratios are similar to those of the normal diet. The polyenoic acid content and the unsaturation index in saturated fat (SF) are minimal, with a virtual absence of linolenic acid. It is of interest to note that the *n*-6 linoleic/*n*-3 linolenic acid ratio in milk lipids, which are the physiologic vehicle of unsaturated fatty acids during postnatal development of mammals, can be calculated in the range of 12.0 and 6.0, respectively, in rabbit and guinea pig milk (14), whereas the (*n*-6)/(*n*-3) ratio is around 3.0 in triglycerides from human milk (15). These values are comparable to those found in the normal diet and in the diets containing olive oil and tallow.

The calculated EFA content in SO is approximately 13% of the calories, in normal diet (N) more than 4%, in OO approximately 1.5%, in T 0.7%, and in SF less than 0.1%, respectively. The optimal requirement of EFA for growth is about 1% of the calories in infants (16) and in the rat (9).

Figure 1 shows the values of the (*n*-9) triene/(*n*-6) tetraene ratios in brain EPG at 1, 20, 40, and 100 days in the five groups of rats. It appears that at 1 day of age the triene/tetraene ratios are low only in the rats born from mothers fed SO or the standard diet, whereas the ratios are higher in the OO group and especially in the T and SF groups. The triene/tetraene ratio is reduced in all groups, even in the T and SF groups at 20 days, and remains low in all groups, except for the SF group, at subsequent age intervals. On the contrary, in the SF group, the ratio increases progressively and linearly in the time period considered. Linearity in the increase of the triene/tetraene ratio in brain EPG during EFA deficiency has been previously observed up to at least 6 months of age (5, 7), i.e., well after the peak of brain growth, and suggests replacement of brain fatty acids at a constant

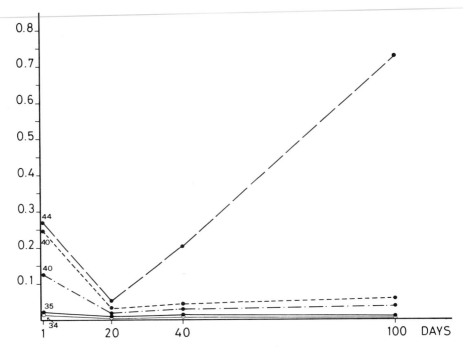

FIG. 1. Ratio of *n*-9 triene/*n*-6 tetraene in brain EPG of rats born from mothers fed the normal diet (○—○) or semisynthetic diets containing 10% w/w of either olive oil (●—·—·—●), sunflower seed oil (●——●), tallow (●- - -●), or saturated fat (●— —●). The young rats were fed the same diets after weaning. Values represent the average of two determinations performed on one large pool of brains at 1 day of age and on three pools of six animals, each at the subsequent age intervals.

rate even when the synthesis of brain polyunsaturated fatty acids *in situ* is considerably reduced (17). Figure 2 shows the values of the *n*-6 pentaene/*n*-3 hexaene ratios in brain EPG in the various groups of rats at different ages. It appears that the values are very high in the SO and in the SF groups, lower in the OO and T group, and very low in the N group. The values in all groups are reduced at 20 days and subsequently remain very low in the N and T groups (receiving diets with a low linoleic/linolenic acid ratio) and slightly higher in the OO group, whereas they return to high levels in the SO and SF groups.

An increase of pentaenes in brain of EFA-deficient animals has been observed by Svennerholm et al. (3), Sun (4), and Crawford and Sinclair (18), as well as in investigations carried out in our laboratory (6, 7). The increase of 22 : 5 *n*-6 pentaenes in brain phospholipids of EFA deficient rats is difficult to explain. Unbalancement of the linoleic/linolenic acid ratio in the trace amount of EFA present in the diet is a possible cause of this effect.

The trends of the *n*-9 triene/*n*-6 tetraene and of the *n*-6 pentaene/*n*-3 hexaene ratios observed in brain EPG of the various groups of animals indicate a high requirement of EFA in a balanced ratio by the central nervous system during the early stages of development, as shown by the appreciable values of the

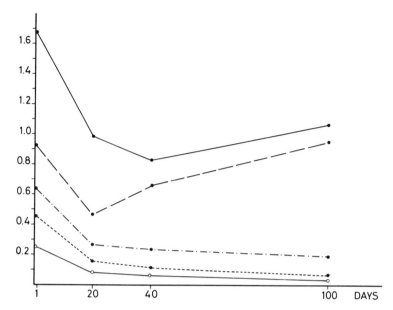

FIG. 2. Ratio *n*-6 pentaene/*n*-3 hexaene in EPG of rats fed semisynthetic diets supplemented with different fats. See Fig. 1 for explanation of symbols, and values.

triene/tetraene ratios at birth even in rats born from mothers fed 10% olive oil in the diet (with a calculated EFA content of 1.6% of the calories), and by the elevated values of the pentaene/hexaene ratios in all animals born from mothers fed diets with a linoleic/linolenic ratio higher than three. This effect may partly depend also upon a low permeability of the placental barrier to fatty acids. The high levels of the pentaene/hexaene ratios observed at birth in brain EPG in several groups of rats fed the semisynthetic diets compared with controls indicate also the importance of a balanced dietary linoleic/linolenic acid ratio.

Reduction of the triene/tetraene ratio and of the pentaene/hexaene ratio during the first 20 days of life results partly from supply of EFA through maternal lactation (19) and derives also from the high rate of deposition of tetraenes and hexaenes in rat brain during this period of time (20).

CONCLUSIONS

The data presented here indicate that brain requires high levels of EFA and also relatively high levels of linolenic acid during fetal development. This is suggested by the elevated triene/tetraene ratios even in brain EPG of animals born from mothers receiving a diet containing 10% olive oil (with a calculated EFA content of 1.6% of the calories) and by an elevation of the pentaene/hexaene ratios in the groups of animal born from mothers fed diets with a linoleic/linolenic acid ratio higher than 3.

Thus, in considering the nutritional value of edible fats in relation to their EFA content, not only the cumulative levels of linoleic and linolenic acid but also the ratio of these essential compounds should be considered. Since the physiologic vehicle of EFA supply during the period of most active postnatal growth of tissues in mammals is maternal milk, the linoleic/linolenic acid ratio detected in this biologic fluid would appear the most adequate for growing tissues. As previously mentioned the linoleic/linolenic acid ratio in milk of most mammals is quite constant and varies between the values of 3 and 6. Our data indicate that both the triene/tetraene and the pentaene/hexaene fatty acid ratios are maintained at the low, normal levels only in animals fed dietary fats with a linoleic/linolenic acid ratio similar to that found in maternal milk. The proposed n-6 pentaene/ n-3 hexaene ratio in tissue lipids may be useful in evaluating the linolenic acid adequacy of the diet, especially in consideration of the relatively low content of this fatty acid in most dietary vegetable oils. It has been recently observed in fact that pathologic symptoms in mice (21) and in monkeys (22) kept on diets with a very high linoleic/linolenic acid ratio are cured by the administration of linolenic acid. These observations suggest an essential role for this unsaturated fatty acid. Among the various vegetable oils used for human consumption, olive oil, widely used in the Mediterranean area, appears to supply an adequate balance of fatty acids of the n-6 and n-3 series, although the total level of EFA is quite lower than that present in most seed oils. This could be an important aspect to be considered in evaluating the nutritional value of olive oil.

ACKNOWLEDGMENTS

This work has been supported by Research Grant N.7300866.11 of the Comitato Tecnologico of the Italian Research Council (CNR).

REFERENCES

1. Mohrhauer, H., and Holman, R. T. (1963): *J. Lipid Res.*, 4:151.
2. Walker, B. (1967): *Lipids*, 2:497.
3. Svennerholm, L., Alling, C., Bruce, A., Karlsson, I., and Sapia, O. (1972): In: *Lipids, Malnutrition and the Developing Brain*, p. 141. Ciba Foundation Symposium, ASP Publishers, Amsterdam.
4. Sun, G. Y. (1972): *J. Lipid Res.*, 13:56.
5. Galli, C., White, H. B., Jr., and Paoletti, R., (1970): *J. Neurochem.*, 17:347.
6. Galli, C., White, H. B., Jr., and Paoletti R. (1971): In: *Chemistry and Brain Development*, p. 425. Plenum Press, New York.
7. Galli, C., White, H. B., Jr., and Paoletti, R. (1971): *Lipids*, 6:378.
8. Galli, C., Trzeciak, H. I., and Paoletti, R. (1971): *Biochim. Biophys. Acta*, 248:449.
9. Holman, R. T. (1960): *J. Nutr.*, 70:405.
10. Galli, C., and Przegalinski, E. (1973): *Pharmacol. Res. Commun.*, 5:239.
11. Galli, C. (1973): *Biochem. Soc. Trans.*, 1 (2):436.
12. Galli, C., Agradi, E., and Paoletti, R. (1974): *Biochim. Biophys. Acta, in press.*
13. Rouser, G., Kritchevsky, G., Galli, C., Yamamoto, A., and Knudson, A. G., Jr. (1966): In: *Inborn Disorders of Sphingolipid Metabolism*, p. 303. Pergamon Press, London.
14. Smith, S., Watts, R., and Dils, R. (1968): *J. Lipid Res.*, 9:52.
15. Crawford, M. A., Sinclair, A. J., Msuya, P. M., and Munhambo, A. (1973): In: *Dietary Lipids and Postnatal Development*, p. 41. Raven Press, New York.

16. Hansen, A. E., Wiese, H. F., Boelsche, A. N., Haggard, M. E., Adam, D. J. D., and Davis, H. (1963): *Pediatrics,* 31 (Suppl. 1, pt. 2):171.
17. Bernsohn, J., and Cohen, S. R. (1972): In: *Lipids, Malnutrition and the Developing Brain,* p. 159. Ciba Foundation Symposium, ASP Publishers, Amsterdam.
18. Crawford, M. A., and Sinclair, A. J. (1972): In: *Lipids, Malnutrition and the Developing Brain,* p. 267. Ciba Foundation Symposium, ASP Publishers, Amsterdam.
19. Galli, C., and Spagnuolo, C. (1974): *Lipids,* 9:1030.
20. Sinclair, A. J., and Crawford, M. A. (1972): *J. Neurochem.,* 19:1753.
21. Rivers, J. P. W., and Davidson, B. C. (1973): *263rd Meeting of the Nutrition Society,* p. 48A, Abstract. December 7, 1973.
22. Sinclair, A. J., Fiennes, R. N. T. W., Hay, A. W. M., Watson, G., and Crawford, M. A. (1973): *263rd Meeting of the Nutrition Society,* p. 49A, Abstract, December 7 (1973).

Lipids, Vol. 1: Biochemistry, edited by R. Paoletti, G. Porcellati, and G. Jacini. Raven Press,
New York © 1976.

Dietary Fat and Susceptibility to Coronary Heart Disease

Flaminio Fidanza

Istituto di Scienza dell'Alimentazione, Università degli Studi, 06100 Perugia, Italy

An important nutritional function of dietary fat is the regulation of blood lipids; this is realized more by the nutrients they contain than by the food component as a whole. Short-term metabolic experiments on humans previously showed that saturated fatty acids increase serum cholesterol, whereas polyunsaturated fatty acids decrease the serum cholesterol level. The equation of Keys et al. (1), $\Delta \text{Chol} = 1.35 \, (2\Delta S - \Delta P)$, in which S is the percentage of energy from saturated fatty acids and P is the percentage of energy from polyunsaturated fatty acids, is a useful well-known approximation. But all the saturated fatty acids do not have the same effect. Lauric, myristic, and palmitic acids probably are the most significant ones, whereas the saturated fatty acids (with less than 12 carbon atoms) and stearic acid are probably comparable to monoenes with no effect.

Attention has not been paid in this equation to transisomers, particularly of oleic acid and to long-chain fatty acids (2). Thus elaidic acid in the presence (but not in the absence) of dietary cholesterol, has a definite effect on increasing serum cholesterol level as compared with oleic acid. In fact elaidic acid was only slightly less active than a mixture of lauric and myristic acid given in equal amounts.

Also erucic acid, which is prevalent in rape seed oil and is now used in some countries, because it is slowly metabolized in some animals including primates, accumulates in muscle, particularly myocardium; it may produce changes in blood lipids, although only preliminary data are available.

The equation derived from international studies of free living population groups has a different expression: $\text{Chol} = 2.85 \, (2\Delta S - \Delta P) + 151.7$. The slope is double, namely the serum cholesterol changes more markedly with variations of dietary fat intake than the metabolic ward experiments indicate (3). This may be due to the presence in the diet of these free living population groups of other components acting as hyper- or hypocholesterolemic agents [e.g., recent reports suggest that onions (4) and garlic (5) have hypocholesterolemic effect].

The mechanisms by which the polyunsaturated fatty acids, and in particular linoleic acid, are hypocholesterolemic agents remains to be established. Some investigators have shown that fecal steroids increase in individuals on high polyunsaturated fat diets and decrease when saturated fat is substituted. According to other workers, reduction of serum cholesterol in response to polyunsaturated fat feeding is attributable to transfer of the sterol to tissue pools, particularly muscle.

It is now well established that hypercholesterolemia is in general associated with precocious severe atherosclerosis and consequently with coronary heart disease (CHD). But besides hypercholesterolemia, two other cardinal risk factors have an additive role in view of their frequency, their impact on risk, and their preventability and reversibility—hypertension and cigarette smoking (6). As we have shown hypercholesterolemia is influenced by diet, and also very likely by hypertension, in terms of both dietary salt and caloric intake. Evidence of these relationships has been obtained through epidemiologic studies. Among them mention will be made here of the International Cooperative Study of Seven Countries (Finland, Greece, Italy, Japan, the Netherlands, the United States, and Yugoslavia).

Sixteen "chunk" samples of men aged 40 to 59, resident for more than 5 years in a defined geographic area, were examined and followed for years, with quinquennial reexaminations. Thus we obtained data on both prevalence and incidence of the disease. The standardized examination procedure included questionnaires on family status and medical history, anthropometry, physical examination, electrocardiogram, blood samples, and qualitative urinalysis (7). The 7-day individual food-weighing technique on a statistical subsample for each cohort, repeated in different seasons, was used for the dietary appraisal (8). For U.S. railroad men, the dietary interview and record was used.

Because the data of the 10-year examinations are not yet complete, we refer here only to the results of the first 5 years, limiting the cohorts to 13 of the 16 samples, since dietary data are scanty for three of them.

The energy intake is generally high because most of the areas are rural, with a preponderance of farmers. The percentage of total energy from protein showed

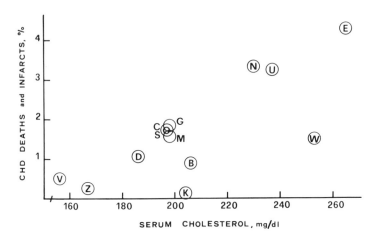

FIG. 1. Five-year incidence rate (age standardized) of CHD deaths and hard criteria nonfatal infarcts plotted against median serum cholesterol concentration of the cohorts, $r = 0.76$. **B,** Belgrade Faculty; **C,** Crevalcore; **D,** Dalmatia; **E,** East Finland; **G,** Corfu; **K,** Crete; **M,** Montegiorgio; **N,** Zutphen; **S,** Slavonia; **U,** U.S. Railroad; **V,** Velika Krsna; **W,** West Finland; **Z,** Zrenjanin.

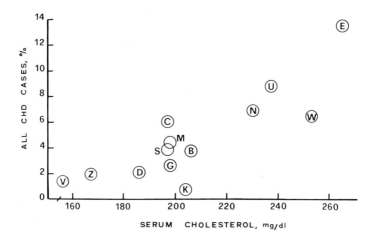

FIG. 2. Incidence rate of all CHD cases plotted against median serum cholesterol concentration of the cohorts, $r = 0.84$. Cohort designations as in Fig. 1.

little variation between populations, although this is not the case for total energy from fats.

Before examining the relationship between some of the dietary data and the incidence of CHD, let us first look at the correlation between one risk factor, serum cholesterol concentration, and incidence rate of CHD. In Fig. 1 the serum cholesterol medians for cohorts are plotted against the 5-year CHD incidence rate (age standardized) of CHD deaths and "hard" criteria infarcts. The correlation coefficient is 0.76. For the incidence rate of all CHD cases, the correlation with serum cholesterol is $r = 0.84$, as shown in Fig. 2.

For dietary energy, expressed per kilogram body weight, no correlation was found with incidence rate of all CHD cases ($r = 0.04$). The same was true for proteins expressed as percentage of total energy ($r = 0.25$). Data on total fats in the diet, as a percentage of total energy, are plotted against the incidence rates of all CHD cases in Fig. 3. The correlation coefficient is rather low ($r = 0.37$). But let us now consider the individual classes of dietary fatty acids. The values expressed as a percentage of total energy available for 12 cohorts are reported in Fig. 4. The saturated fatty acids show major differences among cohorts, rising from only 3% of total energy in Japan to 22% in East Finland.

As shown in Fig. 5, the median concentration of cholesterol in the blood serum was highly correlated with average percentage of energy provided by saturated fatty acids in the diet ($r = 0.88$). Also, as shown in Fig. 6, the incidence rate of all CHD cases proved to be highly correlated with the percentage of energy from saturated fatty acids in the diet ($r = 0.86$).

At this point, mention should be made of the recent findings in primates, particularly rhesus monkeys, receiving peanut oil in cholesterol-containing rations. In spite, of relatively low elevation in blood lipids, they had consistently

ALL CHD CASES, %
CALORIES from FATS, %
E
U
N
W
C
M
S
B
G
D
Z
V
K

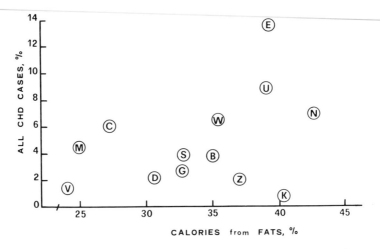

FIG. 3. Incidence rate of all CHD cases plotted against average percentage of total calories from fats in the diet, $r = 0.37$. Cohort designations as in Fig. 1.

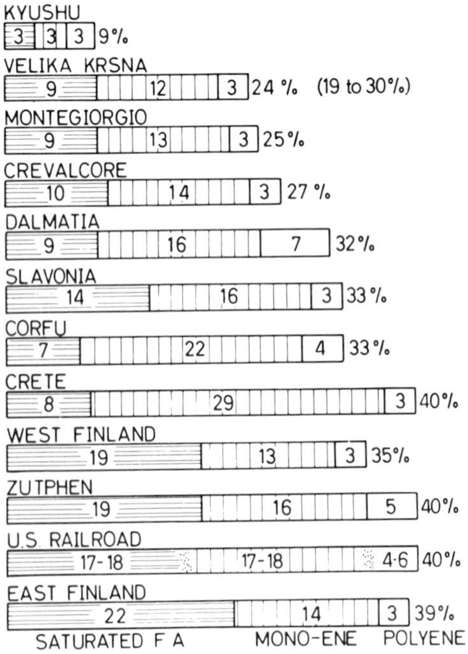

KYUSHU
3 3 3 9%
VELIKA KRSNA
9 12 3 24 % (19 to 30%)
MONTEGIORGIO
9 13 3 25%
CREVALCORE
10 14 3 27 %
DALMATIA
9 16 7 32%
SLAVONIA
14 16 3 33 %
CORFU
7 22 4 33 %
CRETE
8 29 3 40%
WEST FINLAND
19 13 3 35%
ZUTPHEN
19 16 5 40%
U.S. RAILROAD
17-18 17-18 4-6 40%
EAST FINLAND
22 14 3 39%
SATURATED F A
MONO-ENE
POLYENE

FIG. 4. Average percentage of dietary calories provided by saturated, monoene, and polyunsaturated fatty acids.

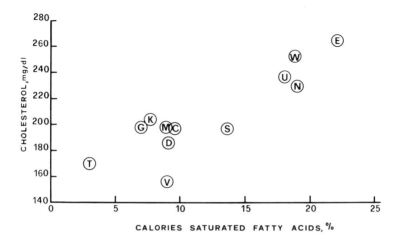

FIG. 5. Median serum cholesterol concentration plotted against average percentage of total calories from saturated fatty acids in the diet, $r = 0.88$. Cohort designations as in Fig. 1.

the most severe arterial intimal cell proliferation and abundant collagen formation with relatively little lipid deposition, the latter being usually located deep in the lesion. The unexpected atherogenicity of peanut oil was related to its content of arachidic and behenic acids and also to the arrangement of the fatty acids in the triglycerides of peanut oil (9).

Going back to the cooperative study of seven countries, for monoene fatty acids, expressed as percentage of total energy, a rather low negative correlation with incidence rate for all CHD ($r = -0.40$) was found; the data are summarized

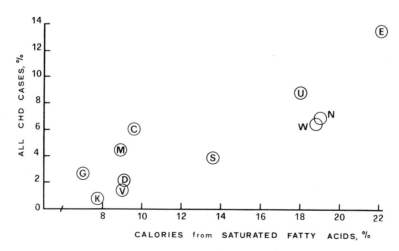

FIG. 6. Incidence rate of all CHD cases plotted against average percentage of total calories from saturated fatty acids in the diet, $r = 0.86$. Cohort designations as in Fig. 1.

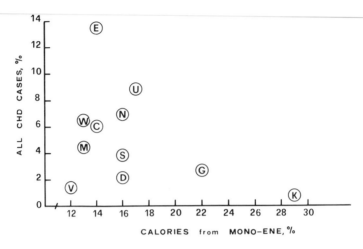

FIG. 7. Incidence rate of all CHD cases plotted against average percentage of total calories from monoene fatty acids in the diet, $r = -0.40$. Cohort designations as in Fig. 1.

in Fig. 7. Serum cholesterol level and dietary monoenes were not correlated. The polyunsaturated fatty acids, as a percentage of energy, showed no correlation with either serum cholesterol medians or CHD incidence. But a high correlation was found between the expression $2S\text{-}P$ and the incidence rate of all CHD cases ($r = 0.83$). The dietary cholesterol intake was available for too few cohorts to correlate with incidence rate of CHD. However, other studies indicate that it does play a role.

Other nonfat nutrients have shown a lower positive or negative correlation with CHD incidence. They have been considered in detail elsewhere (10). The recent results obtained by Kingsbury et al. (11) on a small sample of 80 patients with a proved aortoiliac/femoropopliteal atherosclerosis are interesting. The authors examined their plasma cholesteryl ester fatty acid composition and compared them with the incidence of ischemic heart disease through a 4 year follow-up. Only the cholesteryl-linoleate concentration was significantly different in the patients developing infarcts as compared with those who were not; in fact 70% of the patients with a cholesteryl-linoleate concentration under 35% developed myocardial infarction. Conversely, none of the patients with angina and who had not developed an infarct had values under 35%. A marked inverse relationship was found between oleic and linoleic acid concentration in the cholesteryl esters, without parallel changes in other fatty acids. It is important to note that no significant correlations were found between fatty acid concentration and various other biochemical and rheologic factors, including serum total cholesterol, with the exception of the correlation between linoleic acid and platelet adhesiveness. According to Kingsbury et al. (11) cholesteryl fatty acid analysis rather than total serum cholesterol would be a more sensitive way of determining individual cardiovascular risk status. However, this needs further confirmation.

Another interesting observation from this study is the remarkably high incidence (25% of the patients) of an abnormal fatty acid pattern similar to that found in essential fatty acid (EFA)-deficient animals and children. Such abnormality is characterized by the increase in cholesteryl esters of the specific trienoic acid (20 : 3ω9) by reduced linoleic acid concentrations and by an increase of oleic and palmitoleic acids. These results suggest an increased synthesis of monoenoic acids and monounsaturase activity coupled with a relative inadequacy of linoleic acid. Because this group of patients did not clearly differ from the others with respect to dietary habits, it appears that some unknown factors might have increased their EFA requirements, which could not be met by the low linoleic acid content of the typical British diet. As it is known from animal experiments that saturated and *trans*-monoenoic fatty acids increase EFA requirement, one may speculate on the influence of these factors.

Recently attention has been called to serum triglycerides as possible better predictors of subsequent CHD than serum cholesterol. Still there are technical and environmental factors that can interfere with serum triglyceride determination and results. It is also now clear that dietary fatty acids are much stronger regulators of serum triglycerides than are carbohydrates and, in particular, sucrose.

Mention is now made of some results with primary prevention studies. Most of them were too limited in sample size or not fully controlled. Particularly interesting is the Finnish mental hospital study (12). The designated patients of one hospital served as the experimental group during the first phase, and those in the other hospital served as the control. In the former hospital, the diet was changed so that a large part of milk fat was replaced by soybean oil, given chiefly as filled milk and soft margarine. After the first 6 years of the study, the roles of the hospitals in the experimental design were reversed. The total fat content was 31 to 36% of energy in the control hospital, with milk fat contributing about 17% of energy. The dietary change in the experimental hospital cut the intake of saturated fat to half its original level and increased ingestion of polyunsaturates threefold (linoleic acid 12% of energy). No change in the percentage of energy from total fat occurred in patients in the experimental hospital. At the end of the study the incidence of CHD among the experimental subjects was 61% of that in the controls. In this study also platelet aggregation time was measured using the filtration method of Hornstra (13). This method is based on the use of a pore size, that permits passage of red and white cells and platelets but is occluded by platelet aggregates. Pressure is monitored proximally and distally to the filter, the rise in pressure difference reflecting occlusion of the filter by aggregated platelets. The aggregation time is taken to be the number of seconds required to reach the arbitrarily chosen pressure gradient of 5 mm Hg. The mean aggregation time for the control group was 72 sec as compared to 114 sec for the experimental group; this difference is highly significant. The prolonged aggregation time in the patients receiving a diet low in saturated fat and high in linoleic acid implies decreased aggregability of platelets, which is in good agreement with the decreased tendency toward arterial thrombus formation observed in rats receiving the same diet. It is therefore conceivable that the alteration in platelet function observed in this study contributed to the reduction in CHD mortality in the Helsinki trial. As an explanation of this effect, the authors considered an increased biosynthesis of prostaglandin E_1 (PGE_1), a very potent inhibitor of platelet adhesion and aggregation. Besides this function, PGE_1 has hypotensive effects, antagonizes the hypertensive properties of epinephrine and angiotensin, is a potent inhibitor of catecholamine-induced free fatty acid release from adipose tissue, increases natriuresis especially after its inhibition by vasopressin, and, finally, decreases the influence of both sympathetic and parasympathetic nerve stimulation on myocardial tissue. Thus, the PGs appear to be capable of influencing most of the recognized risk factors in atherosclerotic heart disease. As a working hypothesis, all this suggests, that one possible role of dietary linoleic acid in protecting against CHD is through increased PG synthesis.

Although there are many problems to be solved, there is strong evidence that the amount and the type of fats in the diet are major factors in the etiology of CHD, for which the serum cholesterol level is a key mechanism of this effect.

REFERENCES

1. Keys, A., Anderson, J. T., and Grande, F. (1965): *Metabolism,* 14:747.
2. Vergroesen, A. J. (1972): *Proc. Nutr. Soc.,* 31:323.
3. Fidanza, F. (1970): In: *Nutrition,* edited by J. Musek, K. Osaucova, and Cuthberson, p. 51. Excepta Medica, Amsterdam.
4. Jain, R. C., and Andleigh, H. S. (1969): *Br. Med. J.,* 1:514.
5. Bordia, A., and Bausal, H. C. (1973): *Lancet,* 2:1491.
6. Stamler, J., and Epstein, F. H. (1972): *Prev. Med.,* 1:27.
7. Keys, A. (1970): *Circulation,* 41:Suppl. 1.
8. den Hartog, C., Buzina, R., Fidanza, F., and Roine, P. (1968): *Dietary Studies and Epidemiology of Heart Disease.* Wyt, The Hague.
9. Wissler, R. W., and Vesselinovitch, D. (1974): In: *Nutrition, Vol. 1,* p. 333. Karger, Basel.
10. Fidanza, F. (1973): *Bibl. Nutr. Diet.,* 19:93.
11. Kingsbury, K. J. (1974): Brett, C., Stovold, R., Chapman A., Anderson, J., and Morgan, D. M., *Postgrad. Med. J.,* 50.425.
12. Karvonen, M. J. (1972): *Proc. Nutr. Soc.,* 31:355.
13. Hornstra, G., Lewis, B., Chait, A., Turpeinen, O., Karvonen, M. J., and Vergroesen, A. J. (1973): *Lancet,* 1:1155.

Lipids, Vol. 1: Biochemistry, edited by R. Paoletti, G. Porcellati, and G. Jacini. Raven Press,
New York © 1976.

Bacterial Glycolipids

N. Shaw

*Microbiological Chemistry Research Laboratories, University of Newcastle-upon-Tyne,
Newcastle-upon-Tyne, NE1 7RU, England*

Bacterial glycolipids are now known to be widespread, particularly in gram-
positive bacteria, and are of two types (a) glycosyl diglycerides and (b) acylated
sugar derivatives (1). The most prevalent are the diglycosyl diglycerides of which
nine different types have so far been characterized (2). The disaccharide residue
is glycosidically bound to the 3-position of a *sn*-1,2-diglyceride. Mono-, tri-, and
tetraglycosyl diglycerides have also been isolated, usually in much smaller quantities. The sugar constituents most prevalent are glucose and galactose, although
mannose, glucosamine, glucuronic acid, galacturonic acid, and rhamnose have
all been identified (2). The largest glycolipid of this type is a pentaglycosyl
diglyceride from *Acholeplasma modicum* (3) containing mannoheptose. The latter, although commonly found in the lipopolysaccharides of the gram-negative
cell wall, has not been found previously in glycolipids. Acylated sugars do not
contain glycerol but have the acyl residues esterified directly to the carbohydrate
moiety. The diacyl inositol mannoside from propionic acid bacteria is an interesting example, since it is the first glycolipid to contain inositol (4).

Recent developments have concentrated upon two areas: (a) the structural and
biosynthetic relationship between glycolipids and carbohydrate-containing phospholipids (glycophospholipids) and (b) the recognition of glycolipids or glycophospholipids as important structural components of intracellular polymers.

Phosphorylated derivatives of glycosyl diglycerides have been isolated from a
number of bacteria (5). These phosphoglycolipids are usually glycerophosphoryl
or phosphatidyl derivatives of diglycosyl diglycerides, although derivatives of
monoglycosyl diglycerides are known (5). The most intensively studied are the
phosphoglycolipids for *Streptococci*. In these bacteria both the glycerophosphoryldiglucosyl diglyceride and the phosphatidyldiglucosyl diglyceride co-
occur. However, chemical studies have established (6) that the latter is not a
diacyl derivative of the former. In the glycerophosphoryldiglucosyl diglyceride,
the glycerophosphate residue is linked to the 6-hydroxyl of the terminal glucose
residue, whereas in the phosphatidyldiglucosyl diglyceride the phosphatidyl residue is linked to the 6-hydroxyl of the internal glucose. The glycerophosphate
residues are also of opposite configuration in the two phosphoglycolipids. Diglycosyl diglycerides are known to be the biosynthetic precursors of these phosphoglycolipids, but the source of the additional glycerophosphate or phosphatidyl
residue has not been established despite tentative evidence which suggests that
phosphatidylglycerol may provide both (7). Phosphoglycolipids containing diglucosyl diglyceride (5), digalactosyl diglyceride (8), and galactosylglucosyl di-

glyceride (5) have been described thus far, and it seems probable that similar derivatives of the remaining diglycosyl diglycerides will be discovered.

Phosphatidylinositol mannosides have been known for many years, but the recent discovery of the diacylinositol mannoside suggests (4) a similar relationship to that discussed above for the diglycosyl diglycerides. Propionic acid bacteria also contain a diacyl phosphatidylinositol mannoside, a possible biosynthetic route for which could be the transfer of a phosphatidyl residue to the diacylinositol mannoside, although this reaction has not yet been demonstrated. We have also isolated an acylated inositol derivative from *Corynebacterium acnes,* and although there was no evidence for the presence of an acylated phosphatidylinositol, it is interesting to note the characterization of such a lipid from another species of *Corynebacterium* (9). Glycophospholipids containing glycolipids of both the glycosyl diglyceride and acylated sugar type are therefore important constituents of bacterial lipids.

The presence of lipids as covalently bound constituents of polymers (lipopolysaccharides) present in gram-negative cell walls has been known for many years. The lipid component, lipid A, does not, however, show any structural similarity to the unbound lipids present in the membrane. Recently a number of lipid-containing polymers, lipoglycans, have been isolated from many gram-positive bacteria in which the lipid moiety is clearly recognizable as a glycolipid of either the glycosyl diglyceride or acylated sugar type. Lipoteichoic acids, which have been isolated from many gram-positive bacteria (10), are membrane-associated polymers of glycerophosphate linked to either a diglycosyl diglyceride or a phosphorylated derivative thereof (11). *Micrococcus lysodeikticus* contains a lipomannan (12) consisting of acylated mannose residues. *Thermoplasma acidophilum* also contains a lipomannan (13), but here the lipid residue is a diether analogue of glycosyl diglyceride. Presumably the lipid moieties in these polymers serve to anchor the polymer to the membrane. Thus a relationship between glycolipids, glycophospholipids, and lipoglycans has clearly been established in many bacteria and further developments in this area may be confidently predicted.

REFERENCES

1. Shaw, N. (1970): *Bacteriol. Rev.,* 34:365.
2. Shaw, N. (1975): *Advances in Microbial Physiology, Vol. 12,* p. 141. Academic Press, London.
3. Mayberry, W. R., Smith, P. F., and Langworthy, T. A. (1974): *J. Bacteriol.,* 118:898.
4. Shaw, N., and Dinglinger, F. (1969): *Biochem. J.,* 112:769.
5. Shaw, N., and Stead, A. (1972): *Fed. Eur. Biochem. Soc. Lett.,* 21:249.
6. Fischer, W., Ishizuka, I., Landgraf, H. R., and Herrmann, J. (1973): *Biochim. Biophys. Acta,* 296:527.
7. Pieringer, R. A. (1972): *Biochem. Biophys. Res. Commun.,* 49:502.
8. Veerkamp, J. H., and van Schaik, F. W. (1974): *Biochim. Biophys. Acta,* 348:370.
9. Brennan, P. J. (1968): *Biochem. J.,* 109:158.
10. Coley, J., Duckworth, M., and Baddiley, J. (1972): *J. Gen. Microbiol.,* 73:587.
11. Toon, P., Brown, P. E., and Baddiley, J. (1972): *Biochem. J.,* 127:399.
12. Powell, D., Duckworth, M., and Baddiley, J. (1974): *Fed. Eur. Biochem. Soc. Lett.,* 41:259.
13. Mayberry-Carlson, K. T., Langworthy, T. A., Mayberry, W. R., and Smith, P. F. (1974): *Biochim. Biophys. Acta,* 360:217.

Lipids, Vol. 1: Biochemistry, edited by R. Paoletti, G. Porcellati, and G. Jacini. Raven Press,
New York © 1976.

Newly Discovered Lipids from Streptococci

Werner Fischer

*Institute of Physiological Chemistry, University Erlangen-Nürnberg, D-8520 Erlangen, West
Germany*

Since Macfarlane's pioneer work in 1961 (1), it has become well established
that the membrane lipid composition of gram-positive bacteria differs from that
of gram-negative organisms in having glycolipids that are mostly mono-, di-, and
trihexosyl diglycerides (2). In 1968 diglucosyl diglyceride was found first in
Streptococci to occur not only in a free form but also as a covalently bound
component of phospholipids (3, 4). This led to the discovery of a new lipid class
(5), for which the term phosphoglycolipids was proposed (5, 6).

In this chapter our structural studies on streptococcal phosphoglycolipids (5,
7–9) are summarized; biosynthetic and functional aspects are also discussed.

PROPERTIES AND COMPOSITION

The thin-layer chromatograms (TLCs) in Fig. 1 show the carbohydrate-containing lipids of crude extracts from *S. hemolyticus* D-58 and *S. faecalis,* consisting of mono-, di-, and trihexosyl diglycerides and a series of phosphoglycolipids
(Lipid I–VI[1]) that differ in their abundance and show widely varying polarities.
By comparing crude extracts from *S. faecalis* cells and cytoplasmic membrane,
it was demonstrated that phosphoglycolipids are membrane components as are
all the other polar lipids (9).

Purification is most conveniently performed by column chromatography of the
crude extracts on DEAE cellulose (acetate form), and subsequent separation of
the anionic lipids on silica gel columns (9, W. Fischer, *unpublished data*). Table
1 compiles the molecular composition of the heretofore isolated *Streptococci*
phosphoglycolipids, showing additionally the strains from which they were obtained. The infrared absorption spectra (7, 8) are consistent with the analytic data
and the structure of phosphate ester-containing glycolipids.

STRUCTURE

The lipids listed in Table 1 were subjected to structural analyses, which led
to the formulas given in Fig. 2. The degradation sequences (5, 7–9) are summarized in Fig. 3 and 4, using Lipid VI as an example.

---

[1] In contrast to earlier publications (7–9) and the other chapters in this volume the numbering
of phosphoglycolipids has been changed for structural aspects (cf. Table 1).

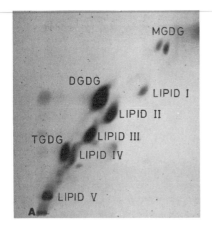 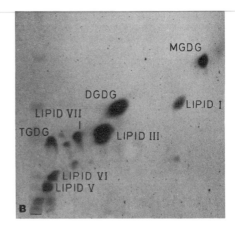

FIG. 1. A: Two-dimensional thin-layer chromatograms of crude lipid extracts from *S. hemolyticus* D-58. **B:** *S. faecalis var. faecalis* on silica gel plates (Merck). The first direction (upward) was run with chloroform–methanol–water (65 : 25 : 4, v/v), and the second direction with chloroform–acetone–methanol–acetic acid–water (50 : 20 : 10 : 10 : 5, v/v). Carbohydrate-containing lipids were visualized with α-naphthol-sulfuric acid. MGDG, Monoglucosyl diglyceride; DGDG, diglucosyl diglyceride; TGDG, triglucosyl diglyceride; Lipids I–VII, phosphoglycolipids. Only Lipids V and VI gave a positive reaction for 1,2-glycols with periodate-Schiff reagent.

TABLE 1. *Analyses of phosphoglycolipids*

Compound	D-Glucose	Glycerol	Acyl groups	Phos-phorus	1,2-Glycol	Reference
			Lipid I			
S. hemolyticus	0.95	1.95	3.79	1.00	0	*
			Lipid II			
S. hemolyticus	2.01	1.99	4.11	1.00	0	*
			Lipid III			
S. hemolyticus	1.95	2.01	4.06	1.00	0	*
S. faec. faec.	1.89	1.97	3.89	1.00	0	(8)
S. faec. zym.	1.97	2.00	4.04	1.00	0	(8)
S. lactis	2.02	2.05	3.92	1.00	0	(5,8)
			Lipid V			
S. hemolyticus	2.10	2.05	2.10	1.00	1.03	(7)
S. faec. faec.	1.98	2.02	2.04	1.00	1.01	(7)
S. faec. zym.	1.98	2.01	1.99	1.00	1.02	(7)
			Lipid VI			
S. faec. faec.	1.83	2.94	3.86	2.00	1.00	(9)
S. faec. zym.	1.92	2.86	4.02	2.00	0.92	(9)

Values are given as mole per mole phosphorus apart from Lipid VI. Values of carbohydrate (anthrone-sulfuric acid method) and of enzymatically determined glucose were identical.
* W. Fischer, *unpublished data.*

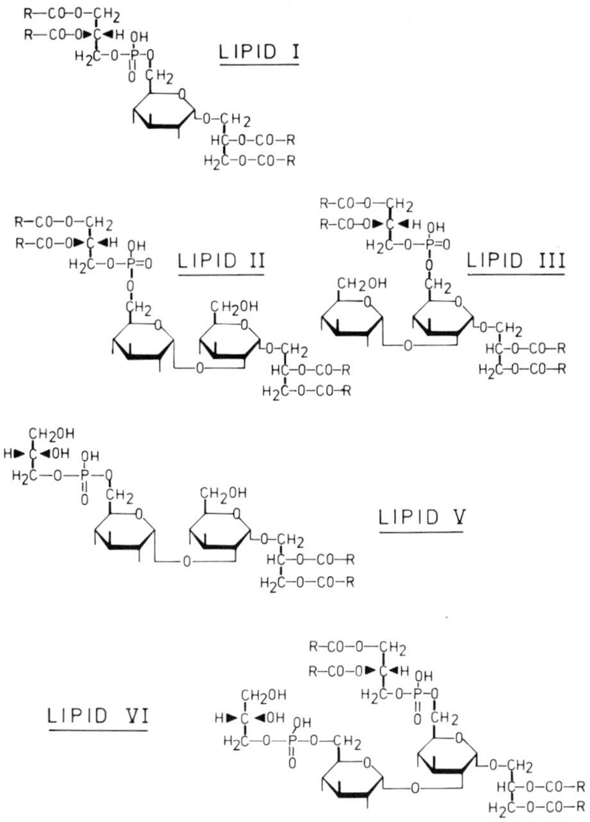

FIG. 2. Structures of *Streptococci* phosphoglycolipids.

Linkage Analysis of the Core

After mild alkaline deacylation the core was analyzed as outlined in Fig. 3. From the results obtained on a two-step acid hydrolysis and strong alkaline degradation with subsequent enzymic breakdown of the alkaline hydrolysis products, it becomes apparent that the core consists of a diglucosylglycerol with two α-glucosidic bonds and two glycerophosphate portions linked to the diglucosylglycerol by diester linkages. By the same degradative sequences the core of Lipid I was shown to be glycerophosphoryl α-glucosylglycerol, whereas the deacylation products of Lipids II, III, and V were identified as glycerophosphoryl diglucosylglycerols. Their basic diglucosylglycerols are chromatographically identical and contain again both glucosyl residues in the α-anomeric form.

Linkage analysis has been achieved by Smith degradation (Fig. 3). The reduced oxidation product was subjected to mild acid hydrolysis, which cleaves readily acetal bonds with negligible attack on the phosphodiester linkages. The break-

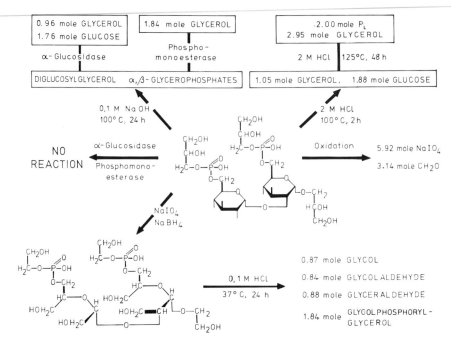

FIG. 3. Degradative sequences establishing the structure of the core of Lipid VI. Quantitative results are given in mole per mole deacylation product, corresponding to 2 moles phosphorus. (For details of Smith degradation and determination of breakdown products, see refs. 7–9.)

down products, containing the carbon atoms adjacent to the linkages of the parent compound were identified and quantitatively determined (7–9). The structure of the phosphorus-containing breakdown product was established after trimethylsilylation by gas-liquid chromatography (GLC) (7) and by mass spectral analysis (9). Using $NaBD_4$ for reduction of the periodate oxidation product, mass spectral analysis revealed additionally that the resulting glycolphosphoryl-glycerol contains two deuterium atoms, namely one in the glycol moiety and the other one at position 1 of the glycerol portion (9). This fact, which indicates that both polyols are derived from aldehyde precursors, together with the formation of about 2 mole proportions glycolphosphorylglycerol proves that in the parent compound both glycerophosphates are α-isomers and that they are attached to C-6 of the glucosyl residues which in turn must have been present in the pyranose form. The glycol liberated as such shows that the third glycerol is α-substituted by a glycosidic bond and consequently the two glycosyl moieties must be linked to a disaccharide. This is confirmed by the glyceraldehyde formed, which proves additionally the linkage in the disaccharide to be a $(1 \rightarrow 2)$ bond. The observed periodate consumption and formaldehyde formation is consistent with this over-all structure of deacylated Lipid VI (Fig. 3).

In the same way, the cores of Lipids II, III, and V were shown to be 1(3)-*O*-

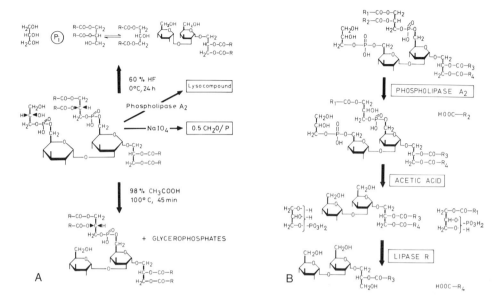

FIG. 4. A: Degradative sequences for the location of acyl groups and the enantiomeric glycerophosphates in Lipid VI. **B:** Degradation of Lipid VI for analysis of the positional distribution of the constituent fatty acids. Lipase R = 1-position specific lipase from *Rhizopus arrhizus*.

[6′(6″)-(α-glycerophosphoryl)-2′-*O*-(hexopyranosyl)-hexopyranosyl]glycerols (7, 8, W. Fischer, *unpublished data*). Obviously, they could be different in the stereochemical configuration and the location of their glycerophosphate moieties. Based on the fact that both glucosyl residues are α-glucosidically linked, α-glucosidase was used for glycerophosphate location. Whereas deacylated Lipid III released 1 mole proportion glucose and glycerophosphoryl glucosylglycerol, the deacylation products of Lipid II and V proved to be resistant. Thus the glycerophosphate is linked to the inner glucose of Lipid III and apparently to the outer one in Lipid II and V (see ref. 7).

Stereochemical Configuration of the Glycerophosphates

The stereochemical configuration of the α-glycerophosphates was accomplished by analyses of the glycerophosphates released on strong alkaline hydrolysis (cf. Fig. 3), whereby phosphodiesters are split via cyclic intermediates resulting in a mixture of α- and β-glycerophosphates. Nevertheless, as can be seen from the data obtained with reference compounds of known stereochemical configuration, the released α-glycerophosphates mainly preserve the configuration they had in the parent compounds (Table 2). Therefore, from the results obtained with phosphoglycolipids (Table 2), we have to conclude that Lipids I, II, and III contain *sn*-glycerol-3-phosphate, Lipid V the unusual *sn*-glycerol-1-phosphate, and in Lipid VI one of the α-glycerophosphates is the *sn*-3-isomer, the other one

TABLE 2. *Composition of glycerophosphates (GP) after alkaline hydrolysis*

Compound	% of phosphorus		
	GP	α-GP	*sn*-3-GP
REFERENCE COMPOUND			
3-*sn*-phosphatidyl-choline	97	39	37
1-*sn*-phosphatidyl-choline	96	40	< 1
3-*sn*-bis(phosphatidic acid)	100	41	40
3-*sn*-phosphatidyl-1'-*sn*-glycerol	98	43	22
PARENT COMPOUND			
Lipid I	97	40	38
Lipid II	96	42	42
Lipid III	101	40	43
Lipid V	95	42	< 1
Lipid VI	92	40	20

For methods see refs. 7–9.

sn-glycerol-1-phosphate. Their distribution among the two glucosyl moieties is shown later in this chapter.

Location of the Acyl Groups

The acyl groups were located by degradation of the native phosphoglycolipids with 60% hydrogen fluoride (w/v), which cleaves rather specifically phosphodiesters and phosphomonoesters (10), releasing the originally diester-linked phosphate group as inorganic phosphate (cf. Fig. 4A). The results summarized in Table 3 led to the overall structures of Lipids I, II, III, and V as they are given in Fig. 2. With Lipid VI one of the glycerophosphates turned out to be not substituted, the other one to be esterified with two fatty acids (Fig. 4A), resulting

TABLE 3. *Main breakdown products obtained from phosphoglycolipids on degradation with 60% hydrogen fluoride*

Lipid I	MGDG, P_i, DG
Lipid II	DGDG, P_i, DG
Lipid III	DGDG, P_i, DG
Lipid V	DGDG, P_i, -, G
Lipid VI	DGDG, P_i DG, G

MGDG = monoglucosyl diglyceride, DGDG = α-kojibiosyl diglyceride, DG = 1,2- and 1,3-diglyceride, G = glycerol in questioning which is the *sn*-3- and the *sn*-1-isomer and how they are distributed among the two glucosyl moieties. Lipid VI was therefore degraded with acetic acid, which is known specifically to hydrolyze phosphodiesters with a neighboring hydroxyl group (11). As expected, besides glycerophosphates a phosphatidyl diglucosyl diglyceride was obtained, which co-chromatographed with Lipid III rather than with Lipid II. Accordingly, by the methods outlined above, it could be definitively shown that this compound carries the *sn*-glycerol-3-phosphate, and that linked to the inner glucose. Consequently the *sn*-1-isomer is the nonacylated glycerophosphate portion and must be attached to the outer glucosyl moiety.

In summary, we can state (cf. Fig. 2) that the basic structural unit of Lipid I is α-glucosyl diglyceride, that of Lipids II, III, V, and VI is α-kojibiosyl diglyceride. Both glycolipids are themselves components of *Streptococci* membranes (3, 4, 12) and can thus be considered to be biosynthetic precursors of phosphoglycolipids, *(vide infra)*. The glycerophosphate substituents are uniformly linked to position 6 of the glucosyl moieties. Lipids I, II, and III are characterized by *sn*-3-phosphatidyl residues, Lipids II and III being positional isomers. Lipid V is striking because of its unusual *sn*-glycerol-1-phosphate substituent. Lipid VI finally contains both the substituents and that just at the same positions as they are found in Lipids III and V, respectively.

BIOSYNTHETIC INTERRELATIONSHIPS

In order to complete the structure the fatty acid composition and their positional distribution in *S. faecalis,* phosphoglycolipids have been studied and compared with those of glyco- and phospholipids. The degradation steps (7, 8) were similar to those used in analysis of Lipid VI which are shown in Fig. 4B. It became apparent that the diglyceride moieties of glyco-, phospho-, and phosphoglycolipids display uniformly not only a similar fatty acid composition but also the same unusual nonrandom distribution pattern (Fig. 5): short-chain fatty acids preferentially linked to position 2 of the glycerol moieties, longer-chain acids accumulated at position 1. The only major differences were varying amounts of C_{19}-cyclopropane and C_{18}-monoenic acid the sums of which, however, turned out to be identical. Since cyclopropane acids have been reported to be formed from the corresponding monoenic acids at the level of preformed lipids (13) the differences found may be caused by varying affinities of the individual lipids to cyclopropane synthetase. The otherwise identical composition and distribution pattern let us suggest that in *S. faecalis* one phosphatidic acid is the common precursor of phospho- and glycolipids.

Thus the diglucosyl diglyceride portions of *S. faecalis* phosphoglycolipids are identical with the membrane α-kojibiosyl diglyceride not only in the cores but also in the fatty acid positioning (Fig. 5). The same is true for the phosphatidyl moieties of Lipids III and VI and those of phosphatidyl- and bisphosphatidyl-glycerol.

These observations supplement the biosynthetic route of Lipid III, which was

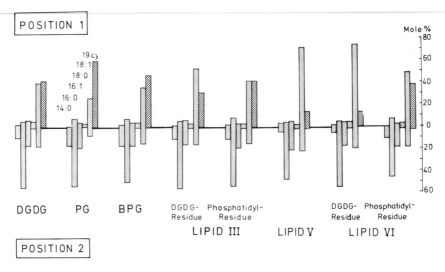

FIG. 5. Positional distribution of fatty acids in glyco-, phospho-, and phosphoglycolipids from *S. faecalis var. faecalis*. DGDG, α-kojibiosyl diglyceride; PG, phosphatidylglycerol; BPG, bis-phosphatidylglycerol, Lipids III, V, and VI, phosphoglycolipids (for structures cf. Fig. 2). cy, C_{19} cyclopropane fatty acid.

worked out by Pieringer (12, 14). By incubating particulate fractions from *S. faecalis* with sn-1,2-di-*O*-acyl-glycerol and uridine diphosphate-glucose (UDP-glucose), Pieringer has demonstrated the consecutive formation of mono- and diglycosyl diglyceride, the latter being then interconverted to Lipid III (12), which is accomplished by a phosphatidyl transfer from either bisphosphatidyl- or phosphatidylglycerol (14). The transferase reacts only with phospholipids from the same organism reflecting an apparent specificity to the fatty acid pattern outlined above. As suggested from the joint occurrence of Lipids III and I, monoglucosyl diglyceride may function as another acceptor.

Our analytic data (Figs. 2 and 5) allow no doubt that Lipids V and VI are also derived from kojibiosyl diglyceride. Thus far, the most probable source of their unusual sn-glycerol-1-phosphate seems to be phosphatidylglycerol. As we found no positional isomers of Lipids III and V in *S. faecalis* (cf. Fig. 1), it can be concluded that the phosphatidyl transferase displays an absolute specificity to position 6 of the internal glucose of the glycolipid acceptor, whereas the glycero-phosphoryl transferase reacts just as specifically with position 6 of the outer glucose (Fig. 6). Consequently the structure of Lipid VI suggests that it might be synthesized by the concerted action of both these transferases. In this connection it might be stressed that the diglucosyl diglycerides of Lipids V and VI are characterized by an identical unusually low content of C_{19}-cyclopropane acid (Fig. 5), which could mean that Lipid VI is derived from Lipid V rather than from Lipid III.

In *S. hemolyticus* D-58 phospho- and glycolipids show differences in fatty acid

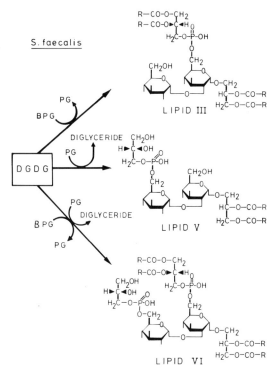

FIG. 6. Biosynthetic pathways of phosphoglycolipids in *S. faecalis* (see text). Abbreviations as in Fig. 5.

composition and positional distribution, but the same distinct patterns are found in the phosphatidyl and glycolipid moieties of Lipids I, II, and III (W. Fischer, *unpublished data*) suggesting a biosynthetic route similar to that in *S. faecalis.* Since, on the other hand, *S. hemolyticus* contains, besides Lipids I and III, Lipid II and probably a phosphatidyltriglucosyl diglyceride (Lipid IV) but lacks Lipid VI (cf. Fig. 1A), its phosphatidyl transferase may have a different acceptor specificity or alternatively several transferases should be present.

OCCURRENCE, QUANTITY, AND FUNCTION

Lipid III was found in four strains of group D *Streptococci (S. faecalis var. faecalis, S. faecalis var. liquefaciens, S. faecalis var. zymogenes, S. faecium)* as well as in *S. lactis* (group N) and in *S. hemolyticus* D-58 (group A), being accompanied in all cases by small amounts of Lipid I. So far Lipid II could be detected only in *S. hemolyticus,* and Lipid VI seems to be confined to group D *Streptococci. S. pyogenes humanus* (group A), *S. dysgalactiae* (group C), and *Streptococci* of groups B, E, P, and Q contain no detectable amount of the heretofore known phosphatidyl glycolipids. All of these strains show, however, small amounts of

TABLE 4. Lipid composition of S. hemolyticus D-58 and S. lactis

Compound	S. hemolyticus	S. lactis
1. Glycolipids		
MGDG	5.1	4.7
DGDG	56.7	7.9
TGDG	6.6	0
2. Aminoacyl-PG	0	12.1
3. Anionic phospholipids		
BPG	16.9	32.7
PG	0.8	21.5
Acyl-PG	0.7	0
4. Phosphoglycolipids		
Lipid I	1.2	trace
Lipid II	4.5	0
Lipid III	4.5	21.0
Lipid IV	—	0
Lipid V	2.8	—
Phosphoglycolipids / Total anionic phospholipids	0.42	0.28

Values, given as mole %, were calculated from the amounts of individual lipids recovered after purification; they are rather too low in the case of phosphoglycolipids for which more purification steps were necessary. Abbreviations as in Figs. 1 and 5.

a compound with the chromatographic behavior and staining properties of Lipid V.

Apart from *Streptococci,* Lipid I has been found in *Pseudomonas diminuta* (15), Lipid II in *Acholeplasma laidlawii* (16), which contains also *sn*-glycero-3-phosphoryl α-kojibiosyl diglyceride (17).

In summary phosphatidyl glycolipids appear to be less widespread than simple glycolipids. But if present, they account for a considerable portion of the anionic lipids (Table 4). There is no doubt that the compounds isolated by us occur in a free state in the cytoplasmic membrane, since the extraction procedures used are not expected to split covalent bonds. Therefore the high amounts of phosphatidyl glycolipids in some *Streptococci* may have as such their own function. Despite the structural similarity, they should not be considered to replace cardiolipin because this lipid is found besides them in even higher amounts.

By contrast, Lipid V and, if present, Lipid VI are as such minor membrane components (Fig. 1, Table 4). On the other hand, recent results suggest that they may occur additionally as covalently bound constituents of lipoteichoic acid, a more complex membrane component of perhaps all gram-positive bacteria (18). Its polyglycerophosphate backbone is thought to be anchored in the cytoplasmic membrane by a covalent bond, probably a phosphodiester, to membrane glycolipids the structure of which has not yet been completely established (18). Evidence of the glycolipid portion of *S. faecalis* lipoteichoic acid being phosphatidyl α-kojibiosyl diglyceride has been afforded by degradation of the polymer with hydrogen fluoride (10), but the linkages of the phosphatidyl group and the teichoic acid to the glycolipid remained unknown. According to recent results obtained with *Staphylococcus aureus* (19) and *Streptococcus sanguis* (20), lipoteichoic acid is synthesized from phosphatidylglycerol rather than from cytidine diphosphate-glycerol (CDP-glycerol), in contrast to wall teichoic acid. If *sn*-glycerol-1-phosphate can be proved to be transferred in this reaction, the *sn*-glycerol-1-phosphate portion would relate Lipids V and VI to lipoteichoic acid, suggesting they could be either starting molecules in biosynthesis or enzymic breakdown products. In favor of the former possibility, one could propose that two transferases may be involved in lipoteichoic acid biosynthesis, a starting enzyme for recognizing the binding site on the glycolipid and a polymerase for chain elongation. If so our studies would have revealed the linkages of the phosphatidyl portion and the polyglycerophosphate to the basic glycolipid of *S. faecalis* lipoteichoic acid. In this context it might be relevant that *S. faecalis* lipoteichoic acid contains, like Lipids V and VI, less C_{19}-cyclopropane acid than the other membrane lipids (10). The restriction of Lipid VI to group D *Streptococci* and the ubiquitous occurrence of Lipid V suggest that the building block of lipoteichoic acid in other than group D *Streptococci* may be *sn*-1-glycerophosphoryl diglycosyl diglyceride. It is noteworthy that a phosphoglycolipid chromatographically similar to Lipid V is also present in *S. aureus* (W. Fischer, *unpublished data*), and that *Lactobacillus bifidum* contains the *sn*-glycerol-1-phosphate derivative of the membrane galactosyl diglyceride (21).

ACKNOWLEDGMENTS

I thank Professor T. Yamakawa and Dr. Ishizuka for leaving the originally joint studies on *S. hemolyticus* D-58 to us and for generous supply of crude lipid extracts. The contribution of my co-workers is gratefully acknowledged. This work was supported by the Deutsche Forschungsgemeinschaft.

REFERENCES

1. Macfarlane, M. G. (1961): *Biochem. J.,* 80:45.
2. Shaw, N., *Bacteriol. Rev.* (1970): 34:365.
3. Ishizuka, I., and Yamakawa, T. (1968): *J. Biochem.,* 64:13.
4. Fischer, W., and Seyferth, W. (1968): *Hoppe Seyler's Z. Physiol. Chem.,* 349:1662.
5. Fischer, W. (1970): *Biochem. Biophys. Res. Commun.,* 41:731.
6. Shaw, N., and Stead, A. (1972): *FEBS Lett.,* 21:249.
7. Fischer, W., Ishizuka, I., Landgraf, H. R., and Herrmann, J. (1973): *Biochim. Biophys. Acta,* 296:527.
8. Fischer, W., Landgraf, H. R., and Herrmann, J. (1973): *Biochim. Biophys. Acta,* 306:353.
9. Fischer, W., and Landgraf, H. R. (1975): *Biochim. Biophys. Acta,* 380:227.
10. Toon, P., Brown, P. E., and Baddiley, J. (1972): *Biochem. J.,* 127:399.
11. Coulon-Morelec, M. J., Faure, M., and Marechal, J. (1960): *Bull. Soc. Chim. Biol.,* 42:867.
12. Pieringer, R. A. (1968): *J. Biol. Chem.,* 243:4894.

13. Zalkin, H., Law, J. H., and Goldfine, H. (1963): *J. Biol. Chem.*, 238:1242.
14. Pieringer, R. A. (1972): *Biochem. Biophys. Res. Commun.*, 49:502.
15. Wilkinson, S. G., and Bell, M. E. (1971): *Biochim. Biophys. Acta*, 248:293.
16. Smith, P. F. (1972): *Biochim. Biophys. Acta*, 280:375.
17. Shaw, N., Smith, P. F., and Verheij, H. M. (1972): *Biochem. J.*, 129:167.
18. Knox, K. W., and Wicken, A. J. (1973): *Bacteriol. Rev.*, 37:215.
19. Glaser, L., and Lindsay, B. (1974): *Biochem. Biophys. Res. Commun.*, 59:1131.
20. Emdur, L. I., and Chiu, T. H. (1974): *Biochem. Biophys. Res. Commun.*, 59:1137.
21. Veerkamp, J. H., and van Schaik, F. W. (1974): *Biochim. Bipophys. Acta*, 348:370.

Lipids, Vol. 1: Biochemistry, edited by R. Paoletti, G. Porcellati, and G. Jacini. Raven Press,
New York © 1976.

The Diphytanyl Glycerol Ether Analogues
of Phospholipids and Glycolipids in Membranes
of *Halobacterium cutirubrum*

M. Kates and S. C. Kushwaha

Department of Biochemistry, University of Ottawa, Ottawa K1N 6N5, Canada

Previous studies of the lipids of extremely halophilic bacteria (1–3) have established the presence in such organisms of a new class of phospholipids and glycolipids, namely derivatives of 2,3-di-*O*-phytanyl-*sn*-glycerol, the phytanyl group
having the configuration $3R,7R,11R,15$-tetramethylhexadecyl (4). The major
phospholipids have been shown to be the diphytanyl glycerol ether analogues of
1-*sn*-phosphatidyl-3'-*sn*-glycero-1'-phosphate (PGP; compound I, Fig. 1; 65% of
total polar lipids), 1-*sn*-phosphatidyl-3'-*sn*-glycerol (PG; compound II, Fig. 1; 4%
of total polar lipids), and 1-*sn*-phosphatidyl-3'-*sn*-glycero-1'-sulfate (PGS; compound III, Fig. 1; 3% of total polar lipids). Note that the configuration of both
glycerol residues in each of these phospholipids is opposite to that of the corresponding glycerols in the analogous diacyl phosphatidylglycerol and derivatives
(5) found in all other organisms. In addition to these phospholipids, the polar lipids of halophilic bacteria also contain an unusual glycolipid sulfate with structure,
2,3-di-*O*-phytanyl-1-*O*-[galactosyl-3'-sulfate-β(1' → 6')-mannosyl-α(1' → 2')-
glucosyl-α(1' → 1)]-*sn*-glycerol (GLS; compound IV, Fig. 1; 25% of total polar
lipids). Small amounts of the triglycosyldiphytanyl ether (TGD; compound V)
derived by desulfation of compound IV are also present.

Studies of Stoeckenius and co-workers (6–9) have established the presence in
Halobacterium halobium of a red membrane, in which the red pigments consist
of C_{50}-carotenoids called bacterioruberins (10, 11), and a purple membrane in
which the chromophore is a retinal-protein complex called bacteriorhodopsin
(8, 9). This purple membrane complex has also been shown to occur in *H.
cutirubrum* (11–13) as well as in other pigmented extremely halophilic bacteria
(14). Under anaerobic conditions in light, the purple membrane appears to function as a light-driven proton pump and the cells utilize the resulting chemiosmotic
gradient for adenosine triphosphate (ATP) synthesis (15, 16). The red membrane
is associated with ATP formation by oxidative phosphorylation under aerobic
conditions in the dark (15, 16).

In view of the different physiologic functions of these two membranes in the
cell, it was of interest to determine the distribution of the major cellular lipids
(compounds I-V) between the red and purple membranes. The present communication deals with a comparison of the lipid composition of the purple and red

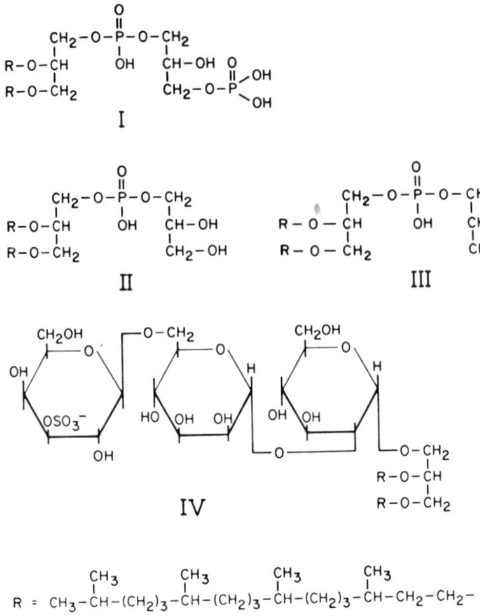

FIG. 1. Structures of major polar lipids in *Halobacterium cutirubrum:* **I**, phosphatidylglycerophosphate (PGP); **II**, phosphatidylglycerol (PG); **III**, phosphatidylglycerosulfate (PGS); **IV**, glycolipid sulfate (GLS). In all compounds the alkyl group R is 3*R*,7*R*,11*R*,15-tetramethylhexadecyl and is linked to glycerol by ether bonds.

membrane from *H. cutirubrum*. More detailed analyses and characterization of these membranes are presented elsewhere (13).

EXPERIMENTAL PROCEDURES

Materials

Deoxyribonuclease (DN-100, from beef pancreas) was purchased from Sigma Chemical Co. Silica gel H for thin-layer chromatography (TLC) was obtained from Brinkmann Instruments Ltd. (Canada), and hexadecyl trimethylammonium bromide (cetyl trimethylammonium bromide, CTAB: technical grade) from J. T. Baker Chemical Co. Authentic sample of *all-trans* retinal was a gift from Dr. O. Isler, Hoffmann-LaRoche and Co. Ltd., Basel, Switzerland. All solvents were glass-distilled before use.

Preparation of Purple and Red Membranes

These membranes were prepared essentially by the procedure of Stoeckenius and coworkers (6–8) from cells grown anaerobically in the light (15, 16) as described in detail elsewhere (13). Briefly, cells of *H. cutirubrum* were first grown aerobically at 37°C in 1.5 liter batches of complex medium for halophiles (17, 18), under fluorescent light at a shaking rate of 120 rpm for 8 hr. Aeration was then reduced by lowering the rate of shaking to 100 rpm for the remainder of the 4-day incubation period; other conditions remained the same. Cells were harvested by centrifugation at 10,000 X g for 20 min, washed twice with basal salt solution (NaCl, 250 g/liter; $MgSO_4$, 9.8 g/liter; KCl, 2 g/liter; pH, 6.5), resuspended in basal salt solution and treated with deoxyribonuclease for 30 min at room temperature. The cell suspension was then dialyzed at 4°C against distilled water for 6 hr and the dialysate was centrifuged at 10,000 X g for 20 min to remove cell debris. The supernatant was centrifuged at 50,000 X g for 1.5 hr, and the pellet was suspended in distilled water and centrifuged at 10,000 X g for 20 min; the pellet was discarded and the supernatant was recentrifuged at 50,000 X g for 1.5 hr. The 50,000 X g pellet ("crude purple membrane") was further purified by centrifugation on a discontinuous sucrose gradient (1.3 M sucrose layered on 1.5 M sucrose) at 40,000 rpm (260,000 X g) for 18 hr in a SW_{41} Ti swinging-bucket rotor in an L_2C 65B Beckman ultracentrifuge at 5°C. The purple membrane band, appearing at the interface of the 1.5 M and 1.3 M sucrose layers and the red membrane band at the top of 1.3 M sucrose layer were collected, dialyzed against distilled water to remove sucrose, and recentrifuged in a similar sucrose gradient as described above. The pure purple and red membranes obtained were stored in sucrose solution at 4°C but were dialyzed against deionized distilled water just before use. Yield of pure purple and red membranes per liter of anaerobic culture was about 6 mg and 2.0 mg, respectively.

For preparation of larger amounts of red membrane, the cells were grown aerobically in the light for 4 days at 37°C in 1-liter batches of complex medium for halophiles at a shaking rate of 120 to 140 rpm and harvested and washed as described above. The red membrane was then prepared as described above, except that the second density gradient centrifugation was done on a discontinuous sucrose gradient of 1 M sucrose layered on 1.3 M sucrose. The pure red membrane formed a single sharp band at the junction of 1.3 M and 1 M sucrose layers; yield 20 mg/liter of aerobic culture.

Extraction of Lipids from Purple or Red Membranes

Total lipids were extracted from the sucrose-free red or purple membranes essentially by the method of Bligh and Dyer (19): to 5 ml of aqueous suspension of the membrane (10 to 20 mg), 12.5 ml of methanol and 6.25 ml of chloroform were added, and the contents were thoroughly mixed. After 5 to 10 min, the protein precipitate was removed by centrifugation; the clear supernatant was diluted with 6.25 ml each of chloroform and water, and the mixture was centrifuged. The chloroform phase was separated, diluted with benzene and brought to dryness on a rotary evaporator; the residual total lipids were dissolved to a known volume in chloroform and aliquots were taken for chemical analyses and chromatography.

Extraction of Retinal from the Purple Membrane

To 400 μl of the sucrose-free dialyzed purple membrane suspension (containing 239 μg of Lowry protein) 100 μl of 0.08 M cetyltrimethyl ammonium bromide (pH, 8) and 300 μl of 0.5 M hydroxylamine base (freshly prepared) were added with mixing. After 1 to 4 hr at room temperature in the dark under nitrogen, the mixture was extracted by adding 2.0 ml methanol and 1.0 ml chloroform, then 1.0 ml each of chloroform and water. After centrifugation the chloroform phase was separated, diluted with benzene, and evaporated to dryness under a gentle stream of nitrogen. The residue was dissolved in ethanol and the amount of retinal oxime was determined spectrophotometrically.

TLC and Identification of Lipid Components

The polar lipid components from purple and red membranes were separated by TLC on silica gel H (500-μm thick layers) in the following solvent systems (2, 3): chloroform/90% acetic acid/methanol (30:20:4;v/v/v) for unidimensional chromatography, and chloroform/methanol/conc NH_4OH (65:35:5;v/v/v) (first direction) and chloroform/90% acetic acid/methanol (30:20:4;v/v/v) (second direction) for two-dimensional chromatography. Before use, TLC plates were washed with chloroform/methanol (1:1,v/v), air dried, and activated at 110°C for 12 hr.

The separated components were detected by charring with sulfuric acid and by specific stains (21): phosphorus spray reagent for phospholipids, and the α-naphthol spray reagent for glycolipids. The phospholipid (1, 2) and glycolipid (3) components were identified as described previously, and were quantitated by scraping each spot directly into digestion tubes and determining the phosphorous or sugar content, respectively.

Chemical Analyses

Phosphorus was determined by a modification (21) of Allen's procedure (22) or by the micromethod of Bartlett (23). Total hexose content was determined by the phenol sulfuric acid procedure of Dubois et al. (24). Protein determinations on the dialyzed membrane suspensions were carried out by the method of Lowry et al. (25).

Measurement of Spectra

Visible and ultraviolet spectra of purple membrane and red membrane suspensions and of retinal, retinyloxime, and red pigments in suitable organic solvents were recorded with a Coleman-Hitachi Perkin-Elmer model 124 spectrophotometer. The amount of each isoprenoid compound was calculated using the following $E_{1cm}^{1\%}$ values: retinal (20), 1,510 at 383 nm in ethanol; retinyloxime (20), 2,020 at 355 nm in ethanol; red pigments (10), 2,540 at 490 nm in acetone.

RESULTS AND DISCUSSION

General Characterization of Membrane Fractions

The highly purified preparation of purple membrane from *H. cutirubrum* showed absorption maxima at 565 nm and 275 nm (13). The ratio of extinction at 565 nm and 275 nm is 1:2, and the molar extinction at 565 nm is 4.8×10^4, essentially as was reported for *H. halobium* purple membrane (8). On disk gel electrophoresis on 7% or 10% polyacrylamide gels, the purple membrane migrated as a single sharp band, the mobility of which corresponded to a protein with a molecular weight of $19.6 \pm 0.8 \times 10^3$ (13); the molecular weight calculated from the amino acid composition (13) was 18.5×10^3, assuming 1 mole of histidine per mole of protein, and from the value determined by sedimentation analysis (13) was 19×10^3. These results provide consistent evidence that the molecular weight of the purple membrane of *H. cutirubrum* is close to 20,000, a value that was significantly lower than that reported (8) for *H. halobium* purple membrane (mw 26,000).

The purified red membrane preparation showed absorption maxima at 535,500 and 470 nm, typical of bacterioruberins (10, 11, 14). On disk gel electrophoresis it showed the presence of at least six bands ranging in molecular weight from 10,000 to 62,000; there was no band in the region of the single protein band from the purple membrane.

The purple membrane of *H. cutirubrum* contained 20% of total lipids by weight and 77% of protein (Table 1). The phosphorus content of the purple membrane was found to be 1.03%, of which only 79% was found in lipids (Table 1), the remainder, presumably, being bound to the protein (ca. 1.8 atoms per mole of protein). The hexose content of the purple membrane was 2.6%, and all of

TABLE 1. *Overall composition of membrane preparations*

Components	Purple	Red
Protein (%)	77	56
Lipid (%)	20	38
Lipid : protein (wt. ratio)	1 : 3.9	1 : 1.5
Retinal[a] (%)	0.48	0
Retinal : protein (mole ratio)	1 : 2.2	—
Bacterioruberins (%)	—	0.15
Total P (%)	1.03	1.8
Total lipid-P (%)	0.81	1.46
Total hexose (%)	2.61	6.9
Total lipid-hexose (%)	2.76	4.57

[a] Extracted in presence of hydroxylamine (see Methods).

the hexose was associated with the lipids (Table 1), indicating that the purple membrane is not a glycoprotein. Extraction of purple membrane in the presence of hexadecyl trimethylammonium bromide and hydroxylamine gave a retinal content of 0.48% corresponding to a retinal : protein mole ratio of 1 : 2.2 (Table 1). In contrast, Oesterhelt and Stoeckenius reported (8) a retinal to protein mole ratio for *H. halobium* purple membrane of 1 : 1.

The lipid content of the red membrane was 38% by weight and the protein accounted for 56% of the weight (Table 1). The phosphorus content was 1.8%, of which 82% was found in the lipids (Table 1), indicating the presence of some phosphoproteins in the membrane. The hexose and hexosamine content of the red membrane was found to be 6.9%, of which 66% was accounted for in the lipids, indicating that the red membrane contains some glycoproteins.

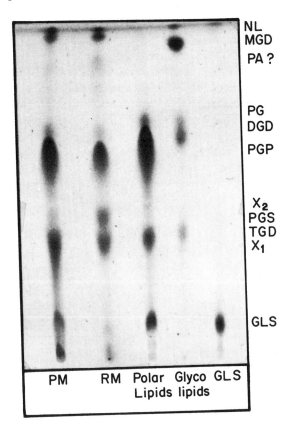

FIG. 2. TLC of total lipids of purple membrane (PM), and red membrane (RM) of *H. cutirubrum.* *Standards:* Total polar lipids of *H. cutirubrum* (1); mixture of glycolipids (2) from *H. cutirubrum;* and pure glycolipid sulfate (GLS) (3). NL, neutral lipids; PA, phosphatidic acid; PG, phosphatidyl-glycerol (II); PGP, phosphatidylglycerophosphate (I); PGS, phosphatidylglycerosulfate (III); MGD, monoglycosyl diether (3); DGD, diglycosyl diether (3); TGD, triglycosyl diether (3); X_1 and X_2, unidentified glycolipids.

Lipid Components of Purple and Red Membranes

TLC of the total lipids from the purple membrane showed that the polar lipids were phosphatidylglycerophosphate, triglycosyl diether, glycolipid sulfate, phosphatidylglycerosulfate, and phosphatidylglycerol (diphytanyl glycerol ether analogues) (Fig. 2, Table 2). In contrast, no sulfated polar lipids (glycolipid sulfate and phosphatidylglycerosulfate) were detected in the red membrane (Fig. 2, Table 2), the main polar lipids being phosphatidylglycerophosphate, phosphatidylglycerol, and two unidentified glycolipids (X_1 and X_2). It is of interest that the contents of the major and minor phospholipids phosphatidylglycerophosphate and phosphatidylglycerol, respectively, are essentially the same in both membranes, and that the glycolipid sulfate and triglycosyl diether in the purple membrane are replaced by the two unidentified glycolipids in the red membrane (Table 2).

The neutral lipids amounted to 7 to 9% of the total lipids in both membranes and consisted mostly of squalene, dihydro- and tetrahydrosqualenes, vitamin MK-8, diphytanyl glycerol ether, and pigments: retinal only in the purple membrane and C_{50}-bacterioruberins only in the red membrane (13).

Thus the striking feature of the purple membrane lipids, relative to those of the red membrane, is the exclusive presence of the sulfated lipids (glycolipid sulfate and phosphatidylglycerosulfate) in the former membrane and their complete absence from the latter (Table 2). Although the precise role of these sulfolipids is not known, it may be noted that the glycolipid sulfate has been found to be essential for formation of stable bilayers of *H. cutirubrum* lipids (26). Thus the lipids in the purple membrane may be involved in maintaining the protein

TABLE 2. *Polar lipid composition of membrane preparations*[a]

Lipid components	Total Lipids[b] (% by weight)	
	Purple membrane	Red membrane
Phospholipids		
Phosphatidylglycerol (II)	4.5	2.7
Phosphatidylglycerophosphate (I)	52.0	52.3
Phosphatidylglycerosulfate (III)	4.8	—
Glycolipids		
X_1	Traces	*
Triglycosyl diether (V)	19.3	Traces
X_2	Traces	*
Glycolipid sulfate (IV)	10.3	Traces

[a] Determined by quantitative two-dimensional TLC (see Methods and ref. 13). See Fig. 1 for structures of polar lipids.

[b] Neutral lipids amounted to 7 to 9% of total lipids in both membranes (13).

* The amount of $X_1 + X_2$ together was about 35% by weight.

component in the specific conformation required for complexing with the retinal. It may also be speculated that the sulfolipids might serve as proton donors for the functioning of the purple membrane as a light-driven proton pump (15, 16).

SUMMARY

The polar lipids of the purple membrane (bacteriorhodopsin) and the red membrane prepared from cells of *H. cutirubrum* were compared by quantitative TLC. The purple membrane contained 77% protein and 20% lipids by weight, the protein component consisting of a single protein moiety, complexed with retinal in mole ratio of 2 : 1, respectively. The red membrane contained 56% protein and 38% lipids including the C_{50}-isoprenoid red pigments (bacterioruberins). The lipids of both membranes contained phosphatidylglycerophosphate (52%) and phosphatidylglycerol (3 to 4%), but the sulfated lipid components, glycolipid sulfate and phosphatidylglycerosulfate, were present *exclusively* in the purple membrane, the red membrane containing instead two unidentified glycolipids. Neutral lipids (squalenes, vitamin MK-8, etc.) were present in both membranes to the extent of 7 to 9%.

ACKNOWLEDGMENT

This work was supported by a grant from the Medical Research Council of Canada.

REFERENCES

1. Kates, M. (1972): In: *Ether Lipids, Chemistry and Biology,* p. 351. Academic Press, New York.
2. Hancock, A. J., and Kates, M. (1973): *J. Lipid Res.,* 14:422.
3. Kates, M., and Deroo, P. W. (1973): *J. Lipid Res.,* 14:438.
4. Kates, M., Joo, C. N., Palameta, B., and Shier, T. (1967): *Biochemistry,* 6:3329.
5. Macfarlane, M. G. (1964): *Adv. Lipid Res.* 2:91.
6. Stoeckenius, W., and Rowen, R. (1967): *J. Cell Biol.,* 34:365.
7. Stoeckenius, W., and Kunau, W. H. (1968): *J. Cell Biol.,* 38:337.
8. Oesterhelt, D., and Stoeckenius, W. (1971): *Nature [New Biol.],* 233:149.
9. Blaurock, A. E., and Stoeckenius, W. (1971): *Nature [New Biol.],* 233:152.
10. Kelly, M., Norgard, S., and Liaaen-Lensen, S. (1970): *Acta Chem. Scand.,* 24:2169.
11. Gochnauer, M. B., Kushwaha, S. C., Kates, M., and Kushner, D. J. (1972): *Ark. Mikrobiol.,* 84:339.
12. Kushwaha, S. C., and Kates, M. (1973): *Biochim. Biophys. Acta,* 316:235.
13. Kushwaha, S. C., Kates, M., and Martin, W. G. (1975): *Can. J. Biochem.,* 53:284.
14. Kushwaha, S. C., Gochnauer, M. B., Kushner, D. J., and Kates, M. (1974): *Can. J. Microbiol.,* 20:241.
15. Oesterhelt, D., and Stoeckenius, W. (1973): *Proc. Natl. Acad. Sci. USA,* 70:2853.
16. Danon, A., and Stoeckenius, W. (1974): *Proc. Natl. Acad. Sci. USA,* 71:1234.
17. Kushwaha, S. C., Pugh, E. L., Kramer, J. K. G., and Kates, M. (1972): *Biochim. Biophys. Acta,* 260:492.
18. Sehgal, S. N., Kates, M., and Gibbons, N. E. (1962): *Can. J. Biochem. Physiol.,* 40:69.
19. Bligh, E. G., and Dyer, W. J. (1959): *Can. J. Biochem. Physiol.,* 37:911.

20. Wald, G., and Brown, P. K. (1954): *J. Gen. Physiol.,* 37:189.
21. Kates, M. (1972): *Techniques of Lipidology,* pp. 354–359. North-Holland, Amsterdam.
22. Allen, R. J. L. (1940): *Biochem. J.,* 34:858.
23. Bartlett, G. R. (1959): *J. Biol. Chem.,* 234:466.
24. Dubois, M., Gilles, K. A., Hamilton, J. K., Rebers, P. A., and Smith, F. (1956): *Anal. Chem.,* 28:350.
25. Lowry, O. H., Rosebrough, N. J., Farr, A. L., and Randall, R. J. (1951): *J. Biol. Chem.,* 193:265.
26. Chen, J. S., Barton, P. G., Brown, D., and Kates, M. (1974): *Biochim. Biophys. Acta.,* 352:202.

Lipids, Vol. 1: Biochemistry, edited by R. Paoletti, G. Porcellati, and G. Jacini. Raven Press,
New York © 1976.

Mechanism of Dehydrogenation of 5α-Cholest-7-en-3β-ol to Cholesta-5,7-dien-3β-ol

Eliahu Caspi and Vangala R. Reddy

Worcester Foundation for Experimental Biology, Shrewsbury, Massachusetts 01545

In this chapter, we discuss what seems to be an obligatory step in the biosynthetic elaboration of cholesterol in rat liver homogenates, namely the transformation of 5α-cholest-7-en-3β-ol (I) to cholesta-5,7-dien-3β-ol (II) (1). This dehydrogenation, which involves the abstraction of the *cis* 5α- and 6α-hydrogen atoms of I (2, 3), occurs in the microsomal fraction of rat livers and requires NAD^+ or NADH and molecular oxygen (4). The NAD^+ and NADH can be replaced by $NADP^+$ or NADPH (4).

From a mechanistic point of view an ionic C-C double bond formation is rationalized in terms of the abstraction of a hydride ion and a proton from neighboring carbon atoms. Several years ago we undertook a study of the mechanism of the dehydrogenation of I to II; we first concerned ourselves with the mode of abstraction of the 6α-hydrogen atom (5). We have synthesized (6α) $[^3H][^{14}C_5]$-5α-cholest-7-en-3β-ol (Ib)($^3H : ^{14}C$ ratio 135) and incubated it with a washed microsomal rat liver preparation in the air (5). The homogenate was prepared with added NAD^+ and *trans*-1,4-bis(2-chlorobenzyl-aminoethyl)cyclohexane dihydrochloride (AY-9944) (5). The AY-9944 inhibits the reduction of the C-7 double bond of II (6, 7). The obtained IIa was essentially devoid of tritium ($^3H : ^{14}C$ ratio 5.02) and the tritium atom abstracted from the C-6α position was found in the water of the medium (5). No tritium was associated with the recovered NADH (5). On this basis we inferred that the 6α-hydrogen (tritium) atom of I apparently is not transferred to the added NAD^+ and is likely abstracted as a proton.

We then turned our attention to the fate of the 5α-hydrogen. Our first objective was to prepare the required (5α)$[^3H]$cholest-7-en-3β-ol (Ic). Oppenauer oxidation of nonradioactive IIa gave III, which was treated with $[^3H]NaBH_4$ (25 mCi) in dry pyridine. The obtained IV was separated from the accompanying (3α)$[^3H]$cholesta-4,7-dien-3β-ol and after extensive purification was oxidized to V. The 5α-cholest-7-en-3-one (V) was equilibrated several times with base to remove exchangeable tritium atoms from C-2 and C-4 and then reduced with $LiAlH_4$ in dry ether to yield Ic. A total of 13 mg of Ic with a specific activity of approximately 7.8 μCi/mg was prepared.

We were now ready to proceed with the incubation experiment. The 5α-tritiated Ic was mixed with a sample of $[^{14}C_5]$I obtained by reduction of a biosyn-

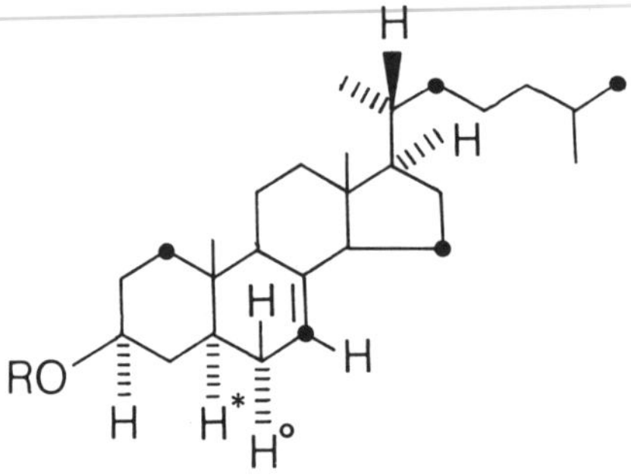

I

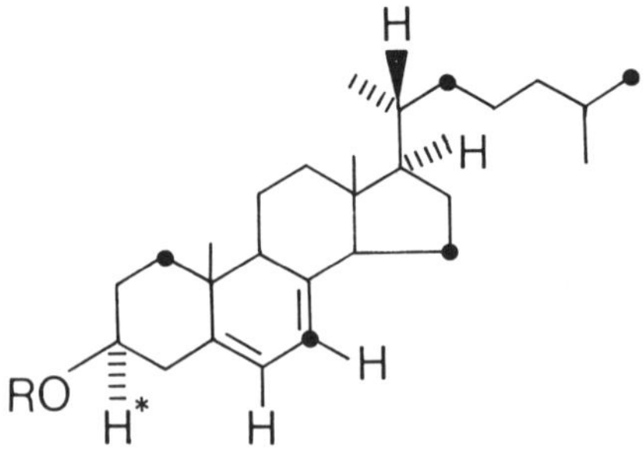

a: H* = H⁰ = ¹H; ● = ¹²C; R = H
b: H* = ¹H; ⁰H = ³H; ● = ¹⁴C; R = H
c: H* = ³H; ⁰H = ¹H; ● = ¹²C; R = H
d: H* = ³H; ⁰H = ¹H; ● = ¹⁴C; R = H
e: H* = ³H; ⁰H = ¹H; ● = ¹⁴C; R = Ac
f: [3α − ³H] − *Ia*

II

a: R = H; H* = ¹H; ● = ¹⁴C
b: R = Ac; H* = ¹H; ● = ¹⁴C
c: R = H; H* = ³H; ● = ¹²C
d: R = Ac; H* = ³H; ● = ¹²C

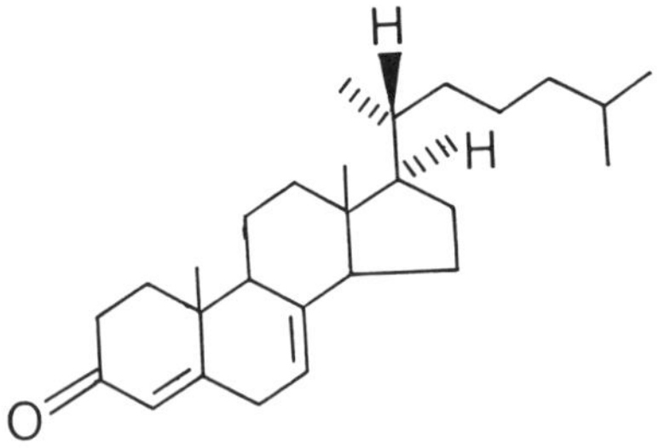

III

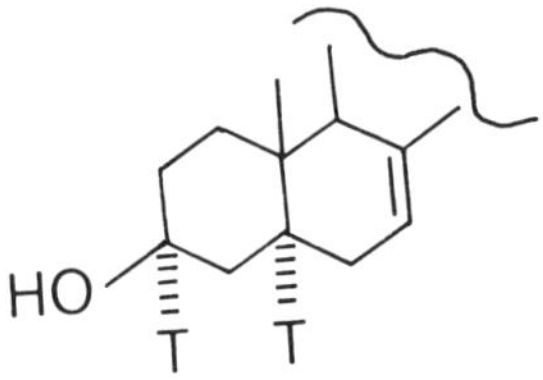

IV

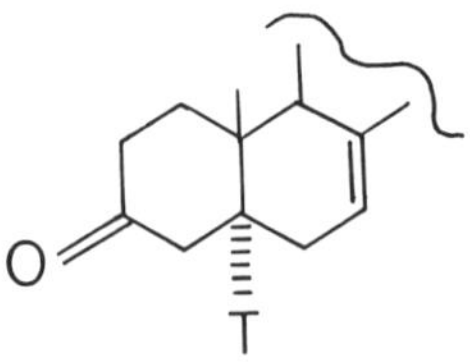

V

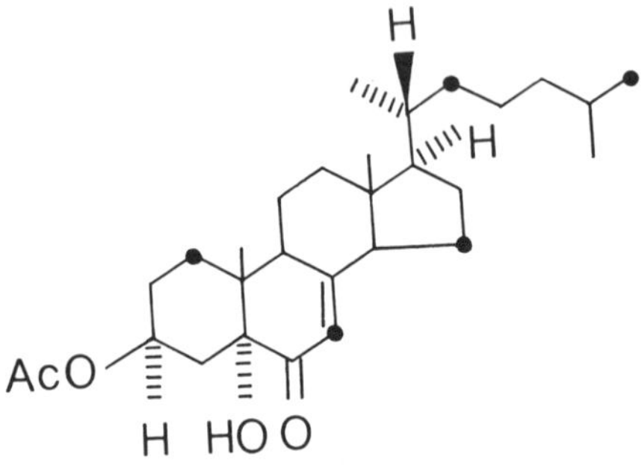

VI

T ≡ ³H

● ≡ ¹⁴C

TABLE 1. *Incubation of [5-³H₁; ¹⁴C₅]-5α-cholest-7-en-3β-ol (Id)ª with rat liver microsomes*

Compounds	[³H] : [¹⁴C] ratio
Id	89.5
Recovered starting material	
(Acetate) Ie	98.7
Cholesta-5,7-dien-3β-ol acetate (IIb)	9.4

For experimental details see ref. (5).

ª 550 μg; 4.3 × 10⁶ dpm of [³H]; [³H] : [¹⁴C] ratio 89.5.

thetic sample of IIa (● = ¹⁴C). The biosynthetic [¹⁴C₅]IIa was prepared by incubating [2-¹⁴C]mevalonic acid with a rat liver homogenate in the presence of the inhibitor AY-9944 (5). The mixed specimen Id (550 μg; 4.3 × 10⁶ dpm of [³H] total; [³H] : [¹⁴C] ratio 89.5) was incubated with rat liver microsomes in the presence of added NAD⁺ and the AY-9944 inhibitor in the air as described (5) for the (6α) [³H] [¹⁴C₅]Ib. The products were recovered, acetylated, and processed until homogeneous. The obtained IIa contained (6.5 × 10³ dpm of [¹⁴C] total; [³H] : [¹⁴C] ratio 9.4) indicating that approximately 13.4% of the Id was converted to IIa (counted as IIb) (Table 1). The dehydrogenation proceeded with the loss of nearly 90% of tritium. The recovered from the incubation mixture (Id; counted as Ie) had a higher [³H] : [¹⁴C] ratio (98.7) revealing that the dehydrogenation involves an isotope effect as previously noted (3, 5).

To determine the location of the tritium atom retained in IIa, the acetate IIb was oxidized with Jones reagent to yield VI ([³H] : [¹⁴C] ratio 2.0). This establishes the distribution of tritium in Ic and hence in Id as follows: ca. 90% at 5α, and 6α positions, ca. 8% at 6β positions, and ca. 2% remained unaccounted.

We have then converted Ie to 5α-Cholestan-3β-ol-7-on acetate VII without loss of tritium. Base equilibration of VII proceeded with the loss of ca. 15% of tritium that was located at C-6. Hence, the distribution of ³H in Ic and Id is as follows: ca. 83% at 5α, ca. 7% at 6α, and ca. 8% at 6β, positions; 2% remains unaccounted for. The small amount of the 6α-tritium will not influence the conclusions, and will therefore be disregarded.

In order to determine the fate of the abstracted 5α-hydrogen (tritium) atom the incubation of Id (600 μg; 5.2 × 10⁶ dpm of [³H], [³H] : [¹⁴C] ratio 89.5) with the washed microsomes was repeated. From the amount of [¹⁴C] (6.9 × 10³ dpm of [¹⁴C]) and the [³H] : [¹⁴C] ratio (9.3) of the obtained IIa (counted as IIb), we estimated that (5.74 × 10⁵ dpm of [³H]) was abstracted in the enzymatic dehydrogenation. Rather surprisingly we found that all the abstracted tritium (6.2 × 10⁵ dpm) was located in the water of the medium and none was associated with the recovered NADH (0.02 × 10⁵ dpm of [³H]). The discrepancy between the calculated amount of tritium which was abstracted and that found in the water is most likely due to the losses incurred in the course of the isolation of IIb.

In view of these results we became concerned whether the washed microsomal preparation indeed requires exogenous NAD⁺ for the dehydrogenation to occur.

To evaluate this point, two incubations of equal amounts of Id with equal aliquots of rat liver microsomes (from the same batch) with and without added NAD^+ were carried out. We observed that the microsomes did not require exogenous NAD^+, since in the absence and in the presence of added NAD^+ (10 mg) the dehydrogenation of Id to IIa (counted as IIb) proceeded in 9% and 11% yield, respectively. We also noted that when $[4-^3H_2]NADH$ was incubated with this microsomal preparation, approximately 55% of the tritium was found in the water indicating a considerable exchange of the isotopic hydrogen of the pyridine nucleotide with the medium (5, 8, 9). Apparently in this microsomal preparation the enzyme(s) mediating the exchange of the C-4 hydrogens of the NADH with the protons of the water were not completely removed.

These observations made the use of the washed microsomal system impractical. Consequently, in further studies we employed a rat liver microsomal acetone powder, which was resuspended in 0.1 M phosphate buffer (pH 7.4). The microsomal powder was prepared by the method of Scallen et al. (10). We first undertook to determine whether the enzyme preparation of the resuspended acetone powder has the capacity to dehydrogenate I to II and whether the process requires exogenous NAD^+. To save the precious $(5\alpha)[^3H]Ic$ these studies were carried out with $(3\alpha)[^3H]-5\alpha$-cholest-7-en-3β-ol (If) which was prepared by reduction of the nonradioactive analogue of V with $[^3H]NaBH_4$.

Incubations of If with the suspended rat liver microsomal acetone powder were carried out in the presence and absence of added NAD^+ under aerobic and anaerobic conditions. From the results summarized in Table 2 it is abundantly clear that the dehydrogenation proceeded only in the air and in the presence of added NAD^+. No conversion of If to IIc (counted as IId) occurred when NAD^+ was not added, in the absence of air or with heat-inactivated enzymes.

We now addressed ourselves to the question of the rate of exchange of tritium from $[4-^3H_2]NADH$ with the medium in the course of the incubation with the resuspended rat liver microsomal acetone powder (Table 3). The results show that little exchange took place in the first 15 min of the incubation and after about

TABLE 2. Cofactors requirement for the conversion of [3α-³H]-5α-cholest-7-en-3β-ol If to [3α-³H] cholesta-5,7-dien-3β-ol (IIc) by a microsomal rat liver acetone powder[a]

Acetone powder (mg)	NAD⁺ (mg)	Gas phase	Conversion[b] (%)
30	10	Air	11.4
30	0	Air	< 0.1
30	10	Helium	0.7
30	10	Nitrogen	0.25
30 mg heat denatured	10	Air	< 0.1

[a] All incubations were carried out using 3.5 μCi of the (3α) [³H]If at 37°C for 3 hr.
[b] Based on the amount of [³H] in the obtained IIc (counted as IId).

TABLE 3. *The rate of exchange of tritium of [4-³H₂]-NADH with protons of the medium under incubation conditions[a]*

Time (min)	Conversion of If to IIc[b] (%)	[³H] exchanged (%)
15	1.1	0.0
30	2.2	17.3
60	3.5	28.5
180	4.7	45.0

[a] The incubations were carried out in air using 20 mg of a rat liver microsomal acetone powder (10) dissolved in 0.1-M phosphate buffer (pH 7.4) (5 ml) to which 10 mg of [4-³H₂]-NADH (1.75 × 10⁵ dpm of [³H]) and 100 μg of the sterol were added.
[b] Isolated and counted as IId.

30 min approximately 20% of the tritium of the added [4-³H₂]NADH was already present in the water. Within the first 30 min of the reaction approximately 2.2% of If was converted to IIc. The observed significant exchange of tritium of [4-³H₂]NADH with the water of the medium precluded extending the incubation time for more than 30 min and thus limited the expected yield of II (Table 3). Consequently, the subsequent experiments with the redissolved rat liver microsomal acetone powder were carried out for 30 min under standardized conditions (see Tables 2 and 3).

At this point we felt it necessary to reevaluate our previous results on the mode of abstraction of the 6α-hydrogen (tritium) atom obtained with the washed microsomes (5). Hence the (6α)[³H] [¹⁴C₅]Ib (1.16 × 10⁵ dpm of [¹⁴C], total; [³H] : [¹⁴C] ratio 61.97) was incubated with a resuspended rat liver microsomal acetone powder to yield IIa (counted as IIb) (2 × 10³ dpm of [¹⁴C]; [³H] : [¹⁴C] ratio 2.11). The calculated amount of tritium, which was abstracted during the dehydrogenation, was (1.2 × 10⁵ dpm of [³H]) all of which (1.3 × 10⁵ dpm of [³H]) was found in the water, and essentially none was associated with the recovered NADH (0.5 × 10³ dpm of [³H]). This therefore confirmed our previous results obtained with the washed microsomal preparation that the 6α-hydrogen (tritium) atom of Ib is not transferred to the added NAD⁺ but is located in the water of the medium.

We now carried out the dehydrogenation experiment of the (5α)-[³H] [¹⁴C₅]Id (7.59 × 10⁴ dpm of [¹⁴C]; [³H] : [¹⁴C] ratio 95.1) with the resuspended rat liver microsomal acetone powder and obtained IIa, (counted as IIb; 1.7 × 10³ dpm of [¹⁴C]; [³H] : [¹⁴C] ratio 10.7). As above we calculated that (1.48 × 10⁵ dpm of [³H]) was abstracted from Ic in the reaction. Again essentially all the tritium was found in the water (1.67 × 10⁵ dpm of [³H]) and the recovered NADH contained only (9.2 × 10³ dpm of [³H]).

Our results confirm the previous observations that the dehydrogenation of I to II is an enzymatic process that takes place in the microsomal fraction of rat livers. The enzyme system requires exogenous NAD^+, molecular oxygen, and involves the abstraction of the *cis* 5α-, and 6α-hydrogen atoms of I. Within the experimental errors of the enzymatic experiments described, our results lead to the conclusion that the 5α and 6α hydrogen (tritium) atoms of I abstracted in the dehydrogenation to II are not transferred to the added NAD^+. It is apparent that both the 5α- and 6α-tritium atoms end up in the water of the medium. The route by which the two hydrogen atoms are transferred to the water of the medium is not yet clear and the problem is currently under investigation.

ACKNOWLEDGMENTS

This work was supported by U.S. Public Health Service Grant AM12156. The generous gift of the AY-9944 from Dr. D. Dvornik, Ayerst Laboratories, Montreal, Canada is gratefully acknowledged.

REFERENCES

1. Frantz, I. D., and Schroepfer, G. J. (1967): *Ann. Rev. Biochem.,* 36:691.
2. Akhtar, M., and Marsh, S. (1967): *Biochem. J.,* 102:462. Dewhurst, S. M., and Akhtar, M. (1967): *Biochem. J.,* 105:1187.
3. Paliokas, A. M., and Schreopfer, G. J. (1968): *J. Biol. Chem.,* 243:453.
4. Scallen, T. J., and Schuster, M. W. (1968): *Steroids,* 12:683.
5. Aberhart, D. J., and Caspi, E. (1971): *J. Biol. Chem.,* 246:1387.
6. Dvornik, D., Kraml, M., Dubuc, J., Givner, M., and Gaudry, R. (1963): *J. Am. Chem. Soc.,* 85:3309.
7. Dvornik, D., Kraml, M., and Bagli, J. F. (1966): *Biochemistry,* 5:1060.
8. Popjak, G., Goodman, D. W. S., Cornforth, J. W., Cornforth, R. H., and Ryhage, R. (1961): *J. Biol. Chem.,* 236:1934.
9. Wilton, D. C., Munday, K. A., Skinner, S. J. M., and Akhtar, M. (1968): *Biochem. J.,* 106:803.
10. Scallen, T. J., Dean, W. J., and Schuster, M. W. (1968): *J. Biol. Chem.,* 243:5202.